Study Guide and Selected Solutions

Organic and Biological
Chemistry
Structures of Life

Karen C. Timberlake

Professor Emeritus,
Los Angeles Valley College

San Francisco • Boston • New York
Capetown • Hong Kong • London • Madrid • Mexico City
Montreal • Munich • Paris • Singapore • Sydney • Tokyo • Toronto

Acquisitions Editor: Maureen Kennedy
Project Editor: Claudia Herman
Managing Editor: Joan Marsh
Cover Designer: Tony Asaro
Cover Photographs: John Bagley, Richard Tauber
Cover Illustration: Blakeley Kim
Manufacturing Coordinator: Vivian McDougal
Marketing Manager: Christy Lawrence

ISBN 0-8053-2986-2

2 3 4 5 6 7 8 9 10–DPC–05 04 03 02 01

www.aw.com/bc

Preface

This Study Guide is intended to accompany *Organic and Biological Chemistry: Structures of Life.* The purpose of this Study Guide is to provide students with additional learning resources to increase their understanding of the key concepts in the text. Each section in the Study Guide is correlated with a chapter in the text. Within each section, there is one or more Learning Exercise that focuses on problem solving, which promotes an understanding of the chemical principles of that Learning Goal. Following the Learning Exercises, a Check List of learning goals and a multiple choice Practice Test provide a review of the entire chapter content. Finally, the Answers and Solutions to Selected Text Problems section gives the odd-numbered solutions for problems at the end of each chapter in the text.

I hope that this Study Guide will help in the learning of chemistry. If you wish to make comments or corrections or ask questions, you can send me an e-mail message to khemist@aol.com.

Karen C. Timberlake
Los Angeles Valley College
Valley Glen, California

To the Student

"One must learn by doing the thing;
though you think you know it, you
have no certainty until you try."
—*Sophocles*

Here you are in a chemistry class with your textbook in front of you. Perhaps you have already been assigned some reading or some problems to do in the book. Looking through the chapter, you may see words, terms, and pictures that are new to you. This may very well be your first experience with a science class like chemistry. At this point you may have some questions about what you can do to learn chemistry. This Study Guide is written with those considerations in mind.

Learning chemistry is similar to learning a new sport such as tennis or skiing or diving. If I asked you how you learn to play tennis or ski or drive a car, you would probably tell me that you would need to practice every day. It is the same with learning chemistry. Learning the chemical ideas and learning to solve the problems depends on the time and effort you invest in it. If you practice every day, you will find that learning chemistry is an exciting experience and a way to understand the current issues of the environment, health, and medicine.

Manage Your Study Time

I often recommend a study system to students in which you read one section of the text and immediately practice the questions and problems that go with it. In this way, you concentrate on a small amount of information and actively use what you learned to answer questions. This helps you to organize and review the information without being overwhelmed by the entire chapter. It is important to understand each section because the sections build like steps. Information presented in each chapter proceeds from the basic to the more complex skills. Perhaps you will only study three or four sections of a chapter. As long as you also practice doing some problems at the same time, the information will stay with you.

Form a Study Group

I highly recommend that you form a study group the first week of your chemistry class. Working with your peers will help you use the language of chemistry. Scheduling a time to meet each week helps you study and prepare to discuss problems. You will be able to teach some things to the other students in the group, and sometimes they will help you understand a topic that puzzles you. You won't always understand a concept right away. Your group will help you see your way through it. Most of all, a study group creates a strong support system whereby students can help each other complete the class successfully.

Go to Office Hours

Finally, go to your tutor's or professor's office hours. Your professor wants you to understand and enjoy learning this material and should have office hours. Often a tutor is assigned to a class or there are tutors available at your college. Don't be intimidated. Going to see a tutor or your professor is one of the best ways to clarify what you need to learn in chemistry.

Using This Study Guide

Now you are ready to sit down and study chemistry. Let's go over some methods that can help you learn chemistry. This Study Guide is written specifically to help you understand and practice the chemical concepts that are presented in your class and in your text. Some of the exercises teach basic skills; others encourage you to extend your scientific curiosity. The following features are part of this Study Guide.

1. Study Goals

The Study Goals give you an overview of what the chapter is about and what you can expect to accomplish when you complete your study and learning of a chapter.

2. Think About It

Each chapter in the Study Guide has a group of questions that encourage you to think about some of the ideas and practical applications of the chemical concepts you are going to study. You may find that you already have knowledge of chemistry in some of the areas. That will be helpful to you. Other questions give you an overview of the chemistry ideas you will be learning.

3. Key Terms

Each chapter in the Study Guide introduces Key Terms. As you complete the description of the Key Terms, you will have an overview of the topics you will be studying in that chapter. Because many of the Key Terms may be new to you, this is an opportunity to determine their meaning.

4. Chapter Sections

Each section of the chapter begins with a list of key concepts to illustrate the important ideas in that section. The summary of concepts is written to guide you through each of the learning activities. When you are ready to begin your study, read the matching section in the textbook and review the sample exercises in the text.

5. Learning Exercises

The Learning Exercises give you an opportunity to practice problem solving related to the chemical principles in the chapter. Each set of Exercises reviews one chemical principle. There is room for you to answer the questions or complete the exercise in this Study Guide. The answers are found immediately following each exercise. (Sometimes they will be located at the top of the next page.) Check your answers right away. If they don't match the answer in the Study Guide, go back to the textbook and review the material again. It is important to make corrections before you go on. In learning tennis, you hit the ball a lot from the baseline before you learn to volley or serve. Chemistry, too, involves a layering of skills such that each one must be understood before the next one can be learned.

At various times, you will notice some essay questions that illustrate one of the concepts. I believe that writing out your ideas is a very important way of learning. If you can put your problem-solving techniques into words, then you understand the patterns of your thinking and you will find that you have to memorize less.

6. Check List

Use the Check List to check your understanding of the Study Goals. This gives you an overview of the major topics in the section. If something does not sound familiar, go back and review. One aspect of being a strong problem-solver is the ability to check your knowledge and understanding as you go along.

7. Practice Test

A Practice Test is found in each chapter. When you have learned the material in a chapter, you can apply your understanding to the Practice Test. If the results of this test indicate that you know the material, you are ready to proceed to the next chapter. If, however, the results indicate further study is needed, you can repeat the learning Exercises in the sections you still need to work on. Answers for all of the questions are included at the end of the Practice Test.

Contents

Contents

11
Introduction to Organic Chemistry

Study Goals

- Identify the number of bonds for carbon and other atoms in organic compounds.
- Describe the tetrahedral shape of carbon with single bonds in organic compounds.
- Classify organic compounds as polar or nonpolar.
- Describe the properties that are characteristic of organic compounds.
- Identify the functional groups in organic compounds.
- Write condensed structural formulas for organic compounds.
- Write structural formulas for constitutional isomers.

Think About It

1. What is the meaning of the term organic?

2. What two elements are found in all organic compounds?

3. In a salad dressing, why is there a layer of vegetable oil floating on the vinegar and water layer?

Key Terms

Match the statements shown below with the following key terms.

a. unsaturated hydrocarbon **b.** constitutional isomer **c.** hydrocarbons
d. alcohols **e.** functional group

1. _____ An atom or group of atoms that influences the chemical reactions of an organic compound
2. _____ A class of organic compounds with one or more hydroxyl (—OH) groups
3. _____ A type of hydrocarbon with one or more carbon–carbon double bonds
4. _____ Organic compounds consisting of only carbon and hydrogen atoms
5. _____ Compounds having the same molecular formula but a different arrangements of atoms

Answers **1.** e **2.** d **3.** a **4.** c **5.** b

11.1 Organic Compounds

- In organic compounds, carbon atoms form four covalent bonds.

—C—C=C—C≡C—

- In organic compounds, H, F, Cl, Br, and I atoms form one covalent bond with carbon, O and S atoms form two covalent bonds, and N atoms form three covalent bonds.

—H —F —Cl —Br —I  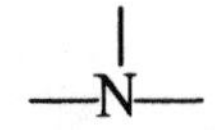

◆ Learning Exercise 11.1

Complete the following structural formulas by adding the correct number of hydrogen atoms:

```
1. C—C—C                         2. C—C—O

                                        Cl
                                        |
3. C—C—N—C                       4. C—C—C—C=C
                                      |
                                      C
```

Answers

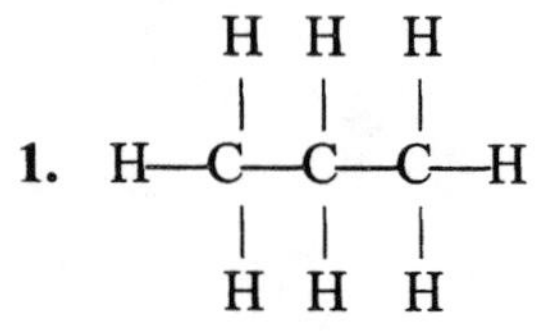

```
                                      H  H
                                      |  |
                                 2. H—C—C—O—H
                                      |  |
                                      H  H

     H  H  H  H                       H  H  Cl H  H
     |  |  |  |                       |  |  |  |  |
3. H—C—C—N—C—H                   4. H—C—C—C—C=C—H
     |  |     |                       |  |  |
     H  H     H                       H  |  H
                                      H—C—H
                                         |
                                         H
```

11.2 The Tetrahedral Shape of Carbon

- Each carbon in an alkane has four bonds arranged so that the bonded atoms are in the corners of a tetrahedron.

◆ Learning Exercise 11.2

Consider the following three-dimensional structure for methane.

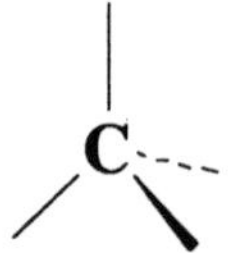

1. Complete the structure by adding hydrogen atoms.

2. What is indicated by the straight lines, wedge, and dashed line used in the three-dimensional structure?

3. What is the shape of this structure?

4. What is the angle between each C—H bond?

Answers

1.

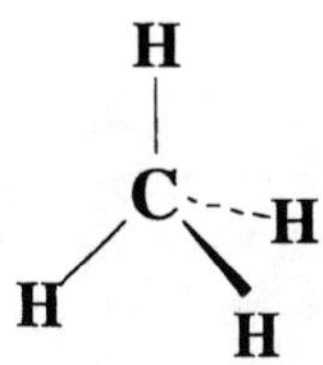

2. The straight lines are in the plane of the page, the wedge comes forward from the page, and the dashed line goes behind the page.
3. Tetrahedron
4. 109.5°

11.3 Polarity of Organic Molecules

- The covalent bonds between carbon and hydrogen atoms are nonpolar.
- The covalent bonds between carbon atom and atoms that are strongly electronegative are polar.
- Symmetrical molecules such as CCl_4, which have polar covalent bonds that cancel, are nonpolar.
- If an organic molecule has polar covalent bonds that do not cancel, the molecule is polar.

◆ Learning Exercise 11.3

Predict if each of the following molecules is polar or nonpolar.

1. CBr_4 __________
2. $CH_3—Br$ __________
3. $CH_3—CH_3$ __________
4. CBr_3Cl __________
5. CH_4 __________
6. $CH_3—CH_2—Cl$ __________

Answers
1. nonpolar; the polar C—Br bonds in the tetrahedral structure cancel out
2. polar
3. nonpolar
4. polar; the C—Cl bond is more polar than the C—Br bonds, which means the polar bonds do not cancel
5. nonpolar
6. polar

11.4 Properties of Organic Compounds

- Organic compounds are compounds of carbon and hydrogen that have covalent bonds, low melting and boiling points, burn vigorously, are nonelectrolytes, and are usually more soluble in nonpolar solvents than in water.

◆ Learning Exercise 11.4

Identify the following as typical of organic (O) or inorganic (I) compounds.

1. ____ have covalent bonds
2. ____ have low boiling points
3. ____ burn in air
4. ____ are soluble in water
5. ____ have high melting points
6. ____ are soluble in nonpolar solvents
7. ____ have ionic bonds
8. ____ form long chains
9. ____ contain carbon
10. ____ are not very combustible
11. ____ have a formula of Na_2SO_4
12. ____ have a formula of $CH_3—CH_2—CH_3$

Answers

1. O	2. O	3. O	4. I	5. I	6. O
7. I	8. O	9. O	10. I	11. I	12. O

11.5 Functional Groups

- The condensed structural formulas are written by grouping the hydrogen atoms with their carbon atoms. For example, CH_3— is a carbon atom bonded to three H atoms.
- Organic compounds are classified by families, which have the same functional groups.
- The *functional group* in an organic molecule is an atom or group of atoms where specific chemical reactions occur.
- Alkenes are hydrocarbons that contain one or more double bonds (C=C); alkynes contain a triple bond (C≡C).
- Alkenes are *unsaturated* because they contain fewer than the maximum number of hydrogen atoms, which can be attached to carbon.

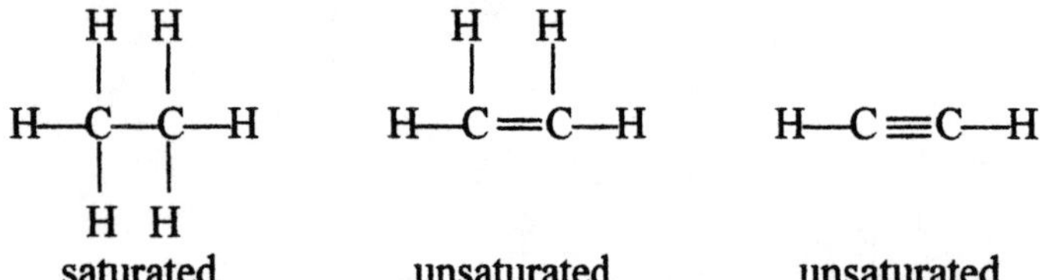

- Alcohols contain a hydroxyl (—OH) group; ethers have an oxygen atom (—O—) between two alkyl groups.
- Aldehydes contain a carbonyl group (C=O) bonded to at least one H atom; ketones contain the carbonyl group bonded to two alkyl groups.
- Carboxylic acids have a carboxyl group attached to hydrogen (—COOH); esters contain the carboxyl groups attached to an alkyl group.
- Amines are derived from ammonia (NH_3), in which alkyl groups replace one or more of the H atoms.

◆ Learning Exercise 11.5A

Write the condensed formulas for the following structural formulas.

```
   H H                              H H H
   | |                              | | |
1. H—C—C—H                       2. H—C—C—C—H
   | |                              | | |
   H H                              H H H
```

```
                                       H
                                       |
                                     H—C—H
   H H H H                          H  |  H H
   | | | |                          |  |  | |
3. H—C—C—C—C—H                   4. H—C—C—C—C—H
   | | | |                          |  |  | |
   H H H H                          H  H  | H
                                       H—C—H
                                         |
                                         H
```

Answers

1. CH_3—CH_3 or CH_3CH_3
2. CH_3—CH_2—CH_3 or $CH_3CH_2CH_3$
3. $CH_3CH_2CH_2CH_3$
4.

```
       CH3
        |
CH3—CH—CH—CH3
        |
       CH3
```

◆ Learning Exercise 11.5B

Classify the organic compounds shown below according to the following functional groups:

a. alkane **b.** alkene **c.** alcohol **d.** ether **e.** aldehyde

1. ____ $CH_3—CH_2—CH{=}CH_2$ **2.** ____ $CH_3—CH_2—CH_3$

3. ____ $CH_3—CH_2—\overset{\overset{\displaystyle O}{\|}}{C}—H$ **4.** ____ $CH_3—CH_2—CH_2—OH$

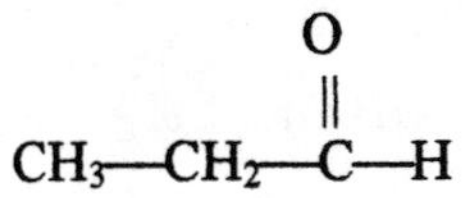

5. ____ $CH_3—CH_2—O—CH_2—CH_3$ **6.** ____ $CH_3—CH_2—CH_2—CH_3$

Answers **1.** b **2.** a **3.** e **4.** c **5.** d **6.** a

◆ Learning Exercise 11.5C

Classify the following compounds by their functional groups.

a. alcohol **b.** aldehyde **c.** ketone **d.** ether **e.** amine

1. ____ $CH_3—CH_2—CH_2—\overset{\overset{\displaystyle O}{\|}}{C}—H$ **2.** ____ $CH_3—CH_2—CH_2—NH_2$

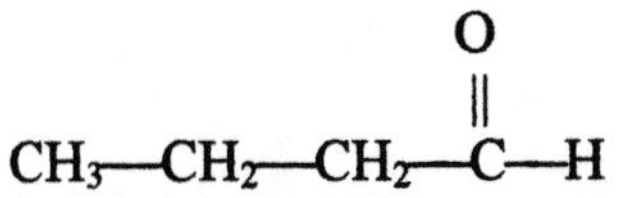

3. ____ $CH_3—CH_2—\overset{\overset{\displaystyle O}{\|}}{C}—CH_2—CH_3$ **4.** ____ $CH_3—CH_2—O—CH_3$

5. ____ $CH_3—\overset{\overset{\displaystyle O}{\|}}{C}—CH_2—CH_3$ **6.** ____ $CH_3—\overset{\overset{\displaystyle O}{\|}}{C}—H$

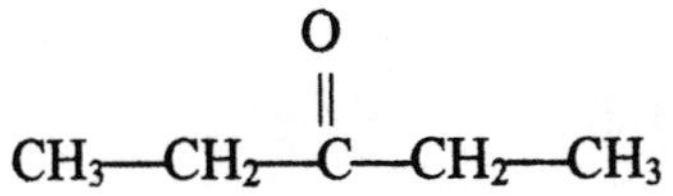

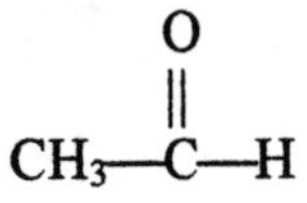

7. ____ $CH_3—CH_2—\overset{\overset{\displaystyle NH_2}{|}}{CH}—CH_3$ **8.** ____ $CH_3—CH_2—\overset{\overset{\displaystyle OH}{|}}{CH}—CH_3$

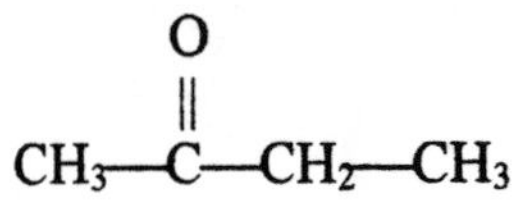

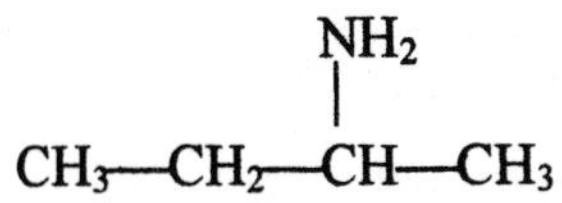

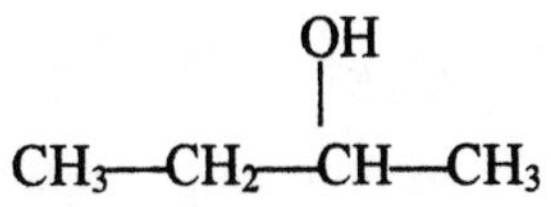

Answers **1.** b **2.** e **3.** c **4.** d
5. c **6.** b **7.** e **8.** a

11.6 Constitutional Isomers

- Constitutional isomers are compounds with the same molecular formula, but a different arrangement of atoms. Examples of constitutional isomers of C_3H_8O are:

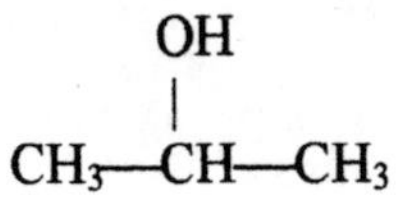

$CH_3—\overset{\overset{\displaystyle OH}{|}}{CH}—CH_3$ $CH_3—CH_2—CH_2—OH$ $CH_3—CH_2—O—CH_3$

◆ Learning Exercise 11.6

Indicate whether each of the following pairs of compounds are constitutional isomers (I), structural formulas of the same compound (S), or structural formulas of different compounds (D).

1. ____ $CH_3—CH_3$ and CH_3 (with CH_3 bonded below: $CH_3—CH_3$)

2. ____ $CH_3—CH_2—OH$ and $CH_3—O—CH_3$

3. ____ $CH_3—CH_2—NH_2$ and $CH_3—N(H)—CH_3$ (H above N)

4. ____ $CH_3—C(=O)—OH$ and $CH_3—O—C(=O)—H$

5. ____ $CH_3—C(=O)—CH_3$ and $CH_3—C(=O)—O—CH_3$

6. ____ $CH_3—CH_2—CH_2—CH_3$ and $CH_3—CH_2—CH_2$ (with CH_3 above the last CH_2)

Answers 1. S 2. I 3. I 4. I 5. D 6. S

Check List for Chapter 11

You are ready to take the practice test for Chapter 11. Be sure that you have accomplished the following learning goals for this chapter. If you are not sure, review the section listed at the end of the goal. Then apply your new skills and understanding to the practice test.

After studying Chapter 11, I can successfully:

____ Identify the number of bonds for carbon and other atoms in organic compounds (11.1).

____ Describe the tetrahedral shape of carbon in carbon compounds (11.2).

____ Predict whether an organic molecule is polar or nonpolar (11.3).

____ Identify properties as characteristic of organic or inorganic compounds (11.4).

____ Identify the functional groups in organic compounds (11.5).

____ Identify the condensed structural formulas for constitutional isomers (11.6).

Practice Test for Chapter 11

Indicate whether each of the following structural formulas are correct (C) or not correct (N).

1. ____
```
H—C—C—H
```

2. ____
```
H  H  H
|  |  |
H—C—C—C—H
|  |  |
H  H  H
```

3. ____
```
  H  H  H
  |  |  |
H—C—O—C—H
  |     |
  H     H
```

4. ____
```
  H  O  H
  |  ‖  |
H—C—C—C—H
  |     |
  H     H
```

5. ____
```
  H  H  H
  |  |  |
H—C—C—N—H
  |  |
  H  H
```

6. ____
```
  H  H  O
  |  |  ‖
H—C=C—C—O—H
```

For problems 7 through 14, indicate whether the following characteristics are typical of (O) organic compounds or (I) inorganic compounds.

7. ____ higher melting points

8. ____ fewer compounds

9. ____ covalent bonds

10. ____ soluble in water

11. ____ ionic bonds

12. ____ combustible

13. ____ low boiling points

14. ____ soluble in nonpolar solvents

Classify the compounds in problems 15 through 22 by the functional groups.

A. alkane **B.** alkene **C.** alcohol **D.** aldehyde
E. ketone **F.** ether **G.** amine

15. ____ $CH_3-CH_2-\overset{O}{\overset{\|}{C}}-CH_3$

16. ____ $CH_3-CH_2-CH_2-OH$

17. ____ $CH_3-CH_2-\overset{CH_3}{\overset{|}{CH}}-CH_3$

18. ____ $CH_3-CH_2-O-CH_3$

19. ____ $CH_3-\overset{NH_2}{\overset{|}{CH}}-CH_2-CH_3$

20. ____ $CH_3-\overset{O}{\overset{\|}{C}}-H$

21. ____ $CH_3-CH_2-\overset{CH_3}{\overset{|}{CH}}-CH_3$

22. ____ $CH_3-CH_2-\overset{OH}{\overset{|}{CH}}-CH_3$

Indicate whether the pairs of compounds in problems 23 through 28 are constitutional isomers (I), structural formulas of the same compound (S), or structural formulas of different compounds (D).

23. ____ $CH_3-CH_2-CH_2-CH_3$ and $CH_3-\overset{CH_3}{\overset{|}{CH}}-CH_3$

24. ____ CH_3-CH_2-OH and $CH_3-\overset{O}{\overset{\|}{C}}-H$

25. ____ $CH_3-CH_2-NH_2$ and $CH_3-\overset{H}{\overset{|}{N}}-CH_2-CH_3$

26 ____ $CH_3-CH_2-\overset{O}{\overset{\|}{C}}-OH$ and $CH_3-\overset{O}{\overset{\|}{C}}-O-CH_3$

27. ____ $CH_3-CH_2-CH_2-CH_3$ and $CH_3-C\equiv C-CH_3$

28. ____ $CH_3-CH=CH-CH_2-OH$ and $CH_3-\overset{O}{\overset{\|}{C}}-CH_2-CH_3$

Answers to Practice Test

1. N	**2.** C	**3.** N	**4.** C	**5.** C
6. C	**7.** I	**8.** I	**9.** O	**10.** I
11. I	**12.** O	**13.** O	**14.** O	**15.** E
16. C	**17.** A	**18.** F	**19.** G	**20.** D
21. A	**22.** C	**23.** I	**24.** D	**25.** D
26. I	**27.** D	**28.** I		

Answers to Selected Problems

11.1 Carbon needs four bonds, nitrogen three bonds, and oxygen two bonds.

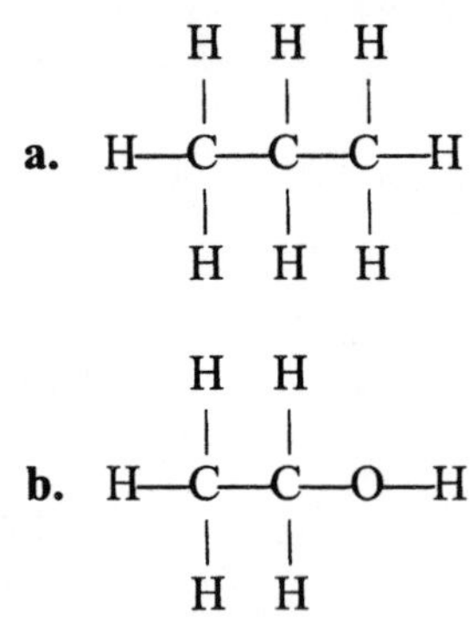

c.

```
    H   H   H
    |   |   |
H—C=C—N—H
```

11.3
a. incorrect; carbon needs four bonds
b. incorrect; hydrogen can have only one bond
c. correct
d. correct

11.5 VSEPR theory predicts that the four bonds in CH_4 will be as far apart as possible, which means that the hydrogen atoms are at the corners of a tetrahedron.

11.7
a. nonpolar
b. nonpolar; four polar bonds cancel in a tetrahedron
c. polar; dipoles do not cancel
d. polar; three nonpolar C—H bonds and one polar bond C—Br

11.9 Organic compounds contain C and H and sometimes O, N, or a halogen atom. Inorganic compounds usually contain elements other than C and H.

a. inorganic **b.** organic **c.** organic
d. inorganic **e.** inorganic **f.** organic

11.11
a. Inorganic compounds are usually soluble in water.
b. Organic compounds have lower boiling points than most inorganic compounds.
c. Organic compounds often burn in air.
d. Inorganic compounds are more likely to be solids at room temperature.

11.13
a. Alcohols contain a hydroxyl group (—OH).
b. Alkenes have carbon–carbon double bonds.
c. Aldehydes contain a C=O bonded to at least one H atom.
d. Esters contain a carboxyl group attached to an alkyl group.

11.15 **a.** Ethers have an —O— group.
b. Alcohols have a —OH group.
c. Ketones have a C=O group between alkyl groups.
d. Carboxylic acids have a —COOH group.
e. Amines contain a N atom.

11.17 **a.** constitutional isomers; same molecular formula but different atom arrangement
b. identical compounds; same order of atoms
c. constitutional isomers; same molecular formula but different atom arrangement
d. constitutional isomers; same molecular formula but different atom arrangement
e. constitutional isomers; same molecular formula but different atom arrangement
f. different compounds; different molecular formulas

11.19 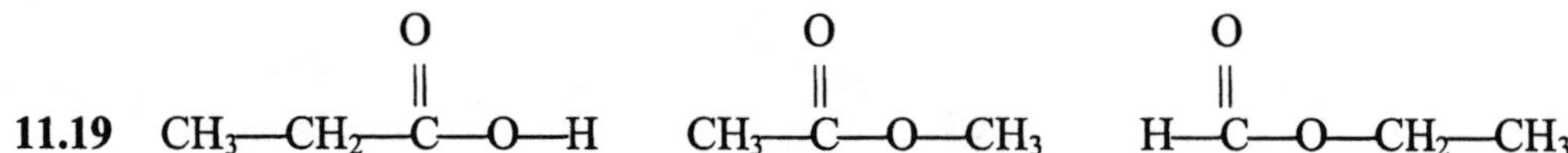

11.21 **a.** Organic compounds have covalent bonds; inorganic compounds have ionic as well as polar covalent and a few have nonpolar covalent bonds.
b. Most organic compounds are insoluble in water; inorganic compounds are soluble in water.
c. Most organic compounds have low melting points; inorganic compounds have high melting points.
d. Most organic compounds are flammable; inorganic compounds are not flammable.

11.23 **a.** butane; organic compounds have low melting points
b. butane; organic compounds burn vigorously in air
c. potassium chloride; inorganic compounds have high melting points
d. potassium chloride; inorganic compounds (ionic) produce ions in water
e. butane; organic compounds are more likely to be gases at room temperature

11.25 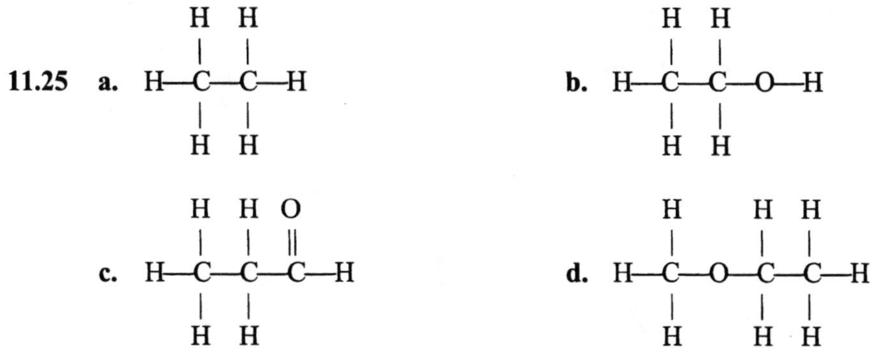

11.27 **a.** A hydroxyl group is —OH; a carbonyl group is C=O
b. An alcohol contains the hydroxyl (—OH) functional group; an ether contains a C—O—C functional group.

c. A carboxylic acid contains the COOH functional group; an ester contains a —COOC— functional group.

11.29 **a.** constitutional isomers **b.** different compounds
c. constitutional isomers **d.** different compounds
e. constitutional isomers **f.** identical compounds
g. different compounds

11.31 **a.** polar **b.** polar **c.** nonpolar
d. polar **e.** polar **f.** polar

11.33 **a.** alcohol **b.** unsaturated hydrocarbon
c. aldehyde **d.** saturated hydrocarbon
e. carboxylic acid **f.** amine
g. tetrahedral **h.** constitutional isomers
i. ester **j.** hydrocarbon
k. ether **l.** unsaturated hydrocarbon
m. functional group **n.** ketone

12
Alkanes

Study Goals

- Draw expanded, condensed, and line-bond structural formulas for alkanes.
- Draw the structural formulas of alkanes and their constitutional isomers.
- Write the IUPAC names for alkanes.
- Write the IUPAC and common names for haloalkanes; draw the condensed structural formulas.
- Write the IUPAC names, and draw the structural formulas for cycloalkanes.
- Describe the physical properties of alkanes and cycloalkanes.
- Write equations for the combustion and halogenation reactions of alkanes and cycloalkanes.

Think About It

1. What type of compound is the octane found in gasoline?

2. What are the products when hydrocarbons are burned in air?

3. Why do oil spills float on water?

Key Terms

Match the statements shown below with the correct key term.

a. alkane **b.** condensed structural formula **c.** main chain
d. combustion **e.** cycloalkane

1. _____ A hydrocarbon that contains only carbon–carbon single bonds

2. _____ An alkane that exists as a cyclic structure

3. _____ The chemical reaction of an alkane and oxygen that yields CO_2, H_2O, and heat

4. _____ The type of formula that shows the arrangement of the carbon atoms grouped with their attached H atoms

5. _____ The longest continuous chain of carbon atoms in a structural formula

Answers **1.** a **2.** e **3.** d **4.** b **5.** c

12.1 Alkanes

- In an IUPAC name, the stem indicates the number of carbon atoms, and the suffix describes the family of the compound. For example, in the name *propane,* the stem *prop* indicates a chain of three carbon atoms and the ending *ane* indicates single bonds (alkane). The names of the first six alkanes follow:

Name	Carbon atoms	Condensed structural formula
Methane	1	CH_4
Ethane	2	$CH_3—CH_3$
Propane	3	$CH_3—CH_2—CH_3$
Butane	4	$CH_3—CH_2—CH_2—CH_3$
Pentane	5	$CH_3—CH_2—CH_2—CH_2—CH_3$
Hexane	6	$CH_3—CH_2—CH_2—CH_2—CH_2—CH_3$

- An expanded structural formula shows a separate line to each bonded atom; a condensed structural formula depicts each carbon atom and its attached hydrogen atoms as a group. The line-bond formula indicates the bonds between carbon atoms. A molecular formula gives the total numbers of atoms.

Expanded Structural Formula	Condensed Structural Formula	Line-bond Formula	Molecular Formula
H H H │ │ │ H—C—C—C—H │ │ │ H H H	$CH_3—CH_2—CH_3$ or $CH_3CH_2CH_3$		C_3H_8

◆ Learning Exercise 12.1A

Indicate if each of the following is a molecular formula (M), an expanded structural formula (E), a condensed structural formula (C), or a line-bond (L) formula.

1. ____ $CH_3—CH_3$
2. ____ C_5H_{12}
3. ____
4. ____
```
  H  H  H
  |  |  |
H—C—C—C—H
  |  |  |
  H  H  H
```
5. ____
```
      CH3
      |
CH3—CH—CH2—CH3
```
6. ____ C_8H_{18}

Answers **1.** C **2.** M **3.** L **4.** E **5.** C **6.** M

◆ Learning Exercise 12.1B

Write the condensed structural formula, line-bond formula, and molecular formula for each of the following expanded structural formulas.

Answers

1. $CH_3—CH_2—CH_2—CH_3$ C_4H_{10}

2. $CH_3—CH(CH_3)—CH_2—CH_2—CH_3$ C_6H_{14}

◆ Learning Exercise 12.1C

Write the condensed structure and name for the straight-chain alkane of each of the following formulas.

1. C_2H_6 ____________________
2. C_3H_8 ____________________
3. C_4H_{10} ____________________
4. C_5H_{12} ____________________
5. C_6H_{14} ____________________

Answers

1. $CH_3—CH_3$, ethane
2. $CH_3—CH_2—CH_3$, propane
3. $CH_3—CH_2—CH_2—CH_3$, butane
4. $CH_3—CH_2—CH_2—CH_2—CH_3$, pentane
5. $CH_3—CH_2—CH_2—CH_2—CH_2—CH_3$, hexane

12.2 IUPAC Naming System for Alkanes

- The IUPAC system is a set of rules used to name organic compounds in a systematic manner.
- Each substituent is numbered and listed alphabetically in front of the name of the longest chain.
- Carbon groups that are substituents are named as alkyl groups or alkyl substituents. An alkyl group is named by replacing the *ane* of the alkane name with yl. For example, $CH_3—$ is named as methyl (from CH_4 methane), and $CH_3—CH_2—$ is named as an ethyl group (from $CH_3—CH_3$ ethane).

Study Note

Example: Write the IUPAC name for the compound

$CH_3—CH_2—CH(CH_3)—CH_3$

Solution: The four-carbon chain *butane* is numbered from the end nearest the side group, which places the *methyl* substituent on carbon 2: *2-methylbutane.*

◆ Learning Exercise 12.2

Provide a correct IUPAC name for each of the following compounds.

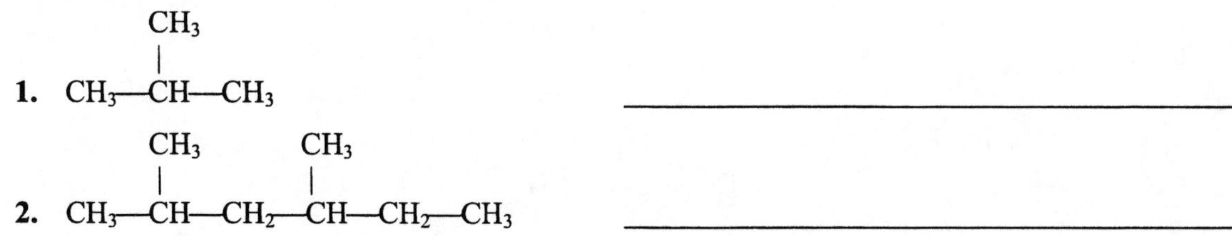

3. $CH_3—CH(CH_3)—CH_2—CH_2—CH(CH_3)—CH_2—CH_3$ ____________________

4. $CH_3—C(CH_3)_2—CH_2—CH_3$ ____________________

5. 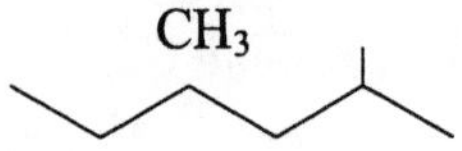 ____________________

Answers
1. 2-methylpropane
2. 2,4-dimethylhexane
3. 2,5-dimethylheptane
4. 2,2-dimethylbutane
5. 2-methylhexane

12.3 Drawing Structural Formulas

- Constitutional isomers have the same molecular formula, but differ in the sequence of atoms in each of their structural formulas.

> **Study Note**
>
> *Example:* Write the constitutional isomers of C_4H_{10}.
>
> *Solution:* Write the first isomer as a main chain of four carbons. Then use a main chain of three carbon atoms and attach the other carbon atom as a substituent.
>
> $CH_3—CH_2—CH_2—CH_3$ and $CH_3—CH(CH_3)—CH_3$

◆ Learning Exercise 12.3A

Write the condensed formula for each of the following compounds:

1. hexane
2. methane
3. 2,4-dimethylpentane
4. propane

Answers

1. $CH_3—CH_2—CH_2—CH_2—CH_2—CH_3$
2. CH_4
3. $CH_3—CH(CH_3)—CH_2—CH(CH_3)—CH_3$
4. $CH_3—CH_2—CH_3$

◆ Learning Exercise 12.3B

Write the condensed structural formula for each of the following alkanes.

1. pentane

2. 2-methylpentane

3. 4-ethyl-2-methylhexane

4. 2,2,4-trimethylhexane

Answers

1. $CH_3—CH_2—CH_2—CH_3—CH_3$

2. $CH_3—\overset{\displaystyle CH_3 \atop |}{CH}—CH_2—CH_3—CH_3$

3. $CH_3—\overset{\displaystyle CH_3 \atop |}{CH}—CH_2—\overset{\displaystyle CH_2—CH_3 \atop |}{CH}—CH_2—CH_3$

4. $CH_3—\underset{\displaystyle | \atop CH_3}{\overset{\displaystyle CH_3 \atop |}{C}}—CH_2—\overset{\displaystyle CH_3 \atop |}{CH}—CH_2—CH_3$

◆ Learning Exercise 12.3C

Write the line-bond structural formula for each of the following compounds.

1 pentane

2. 2-methylbutane

3. 2,4-dimethylhexane

4. 2,3,4-trimethylpentane

Answers

a.

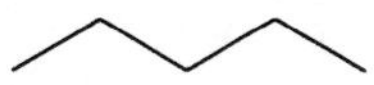

b.

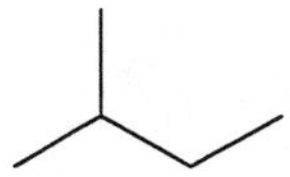

c.

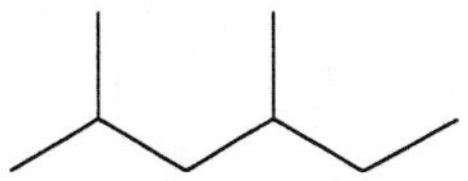

d.

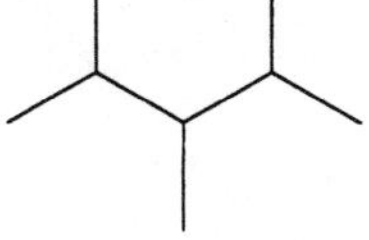

◆ Learning Exercise 12.3D

Explain how it is possible to have two different compounds with the molecular formula C_4H_{10}.

Answer If the atoms can be arranged in two or more different patterns, the different structures are called structural isomers. For C_4H_{10}, the four carbon atoms may be arranged in a straight chain, or they may be arranged to give a chain of three carbon atoms with a one-carbon side group.

◆ Learning Exercise 12.3E

Write the condensed formulas and names for all the isomers with the following molecular formulas.

1. C_4H_{10} (two isomers)

2. C_5H_{12} (three isomers)

3. C_6H_{14} (five isomers)

Answers

1. $CH_3—CH_2—CH_2—CH_3$
butane

CH_3
|
$CH_3—CH—CH_3$
2-methylpropane

2. $CH_3—CH_2—CH_2—CH_2—CH_3$
Pentane

CH_3
|
$CH_3—CH—CH_2—CH_3$
2-methylbutane

CH_3
|
$CH_3—C—CH_3$
|
CH_3
2,2-dimethylpropane

3. $CH_3—CH_2—CH_2—CH_2—CH_2—CH_3$
Hexane

CH_3
|
$CH_3—CH—CH_2—CH_2—CH_3$
2-methylpentane

CH_3
|
$CH_3—CH_2—CH—CH_2—CH_3$
3-methylpentane

CH_3
|
$CH_3—C—CH_2—CH_2$
|
CH_3
2,2-dimethylbutane

CH_3
|
$CH_3—CH—CH—CH_3$
|
CH_3
2,3-dimethylbutane

12.4 Haloalkanes

- In a haloalkane, a halogen atom, F, Cl, Br, or I replace a hydrogen atom in an alkane.
- A halogen atom is named as a substituent (fluoro, chloro, bromo, iodo) attached to the alkane chain.

◆ Learning Exercise 12.4A

Write a correct IUPAC (or common name) for the following.

1. $CH_3—CH_2—Br$

2. 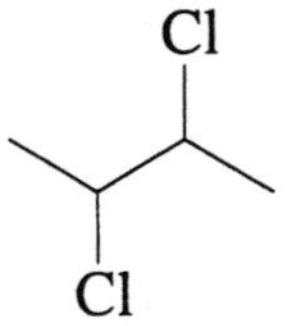 (skeletal structure with Cl on carbon 2 and Cl on carbon 3 of a four-carbon chain)

3. $CH_3—CH_2—C(Cl)(Cl)—CH_2—CH_3$

4. $CH_3—CH_2—CH(Cl)—CH_2—CH(Br)—CH_3$

5. $CH_3—CH_2—CH_2—CH(F)—Cl$

Answers **1.** bromoethane (ethyl bromide) **2.** 2,3-dichlorobutane **3.** 3,3-dichloropentane **4.** 2-bromo-4-chlorohexane **5.** 1-chloro-1-fluorobutane

◆ Learning Exercise 12.4B

Write the condensed formula for each of the following haloalkanes.

1. ethyl chloride

2. bromomethane

3. 3-bromo-1-chloropentane

4. 1,1-dichlorohexane

5. 2,2,3-trichlorobutane

6. 2,4-dibromo-2,4-dichloropentane

Answers

1. $CH_3—CH_2—Cl$

2. $CH_3—Br$

3. 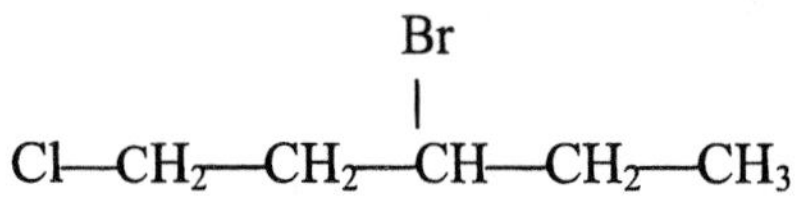
$$\begin{array}{c} \quad\quad\quad\quad\quad\quad\quad\quad\; Br \\ \quad\quad\quad\quad\quad\quad\quad\quad\; | \\ Cl—CH_2—CH_2—CH—CH_2—CH_3 \end{array}$$

4. 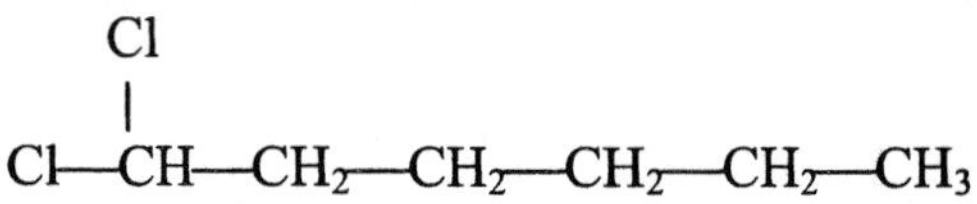
$$\begin{array}{c} Cl \quad\quad\quad\quad\quad\quad\quad\quad\quad\quad\quad\quad\quad\quad\quad \\ | \quad\quad\quad\quad\quad\quad\quad\quad\quad\quad\quad\quad\quad\quad\quad \\ Cl—CH—CH_2—CH_2—CH_2—CH_2—CH_3 \end{array}$$

5. 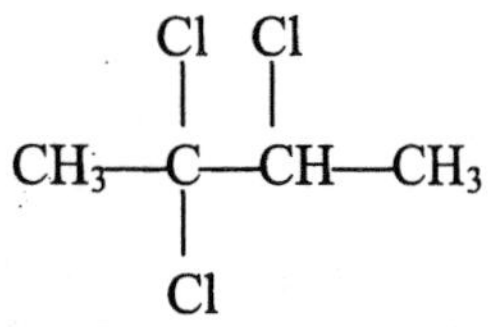
$$\begin{array}{c} \quad\; Cl \;\; Cl \quad\quad \\ \quad\; | \quad\; | \quad\quad \\ CH_3—C—CH—CH_3 \\ \quad\; | \quad\quad\quad\quad \\ \quad\; Cl \quad\quad\quad\quad \end{array}$$

6. 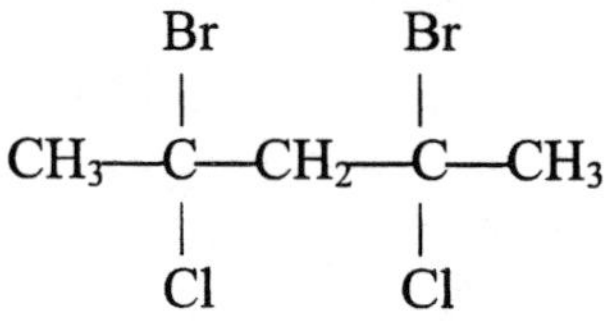
$$\begin{array}{c} \quad Br \quad\quad\quad\; Br \quad \\ \quad | \quad\quad\quad\quad | \quad \\ CH_3—C—CH_2—C—CH_3 \\ \quad | \quad\quad\quad\quad | \quad \\ \quad Cl \quad\quad\quad\; Cl \quad \end{array}$$

12.5 Cycloalkanes

- In cycloalkanes, the carbon atoms form a cyclic structure.
- Cycloalkanes are named by placing the prefix *cyclo* in front of the name of the alkane with the same number of carbon atoms.
- In cycloalkanes, branches are named as alkyl groups and numbered if there are two or more on a ring.

◆ Learning Exercise 12.5A

Write the IUPAC name for each of the following cycloalkanes.

1.  (cyclopropane ring)
2. 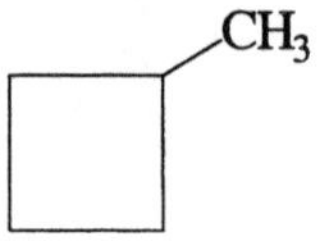 (cyclobutane ring with CH_3)
3. (cyclohexane ring with CH_3 and CH_3 on adjacent carbons)
4. 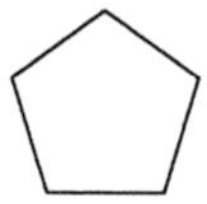 (cyclopentane ring)
5. 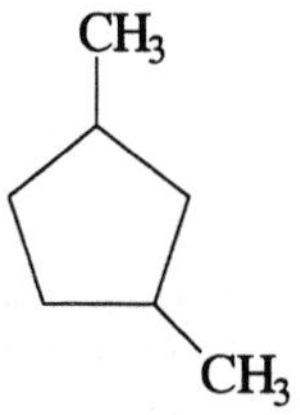 (cyclopentane ring with CH_3 and CH_3)

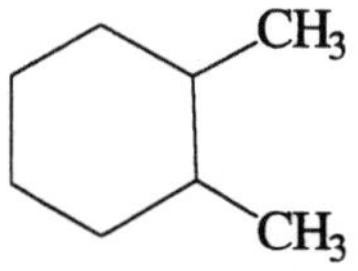

Answers

1. cyclopropane **2.** methylcyclobutane **3.** 1,2-dimethylcyclohexane
4. cyclopentane **5.** 1,3-dimethylcyclopentane

◆ Learning Exercise 12.5B

Write the condensed structural formula for each of the following cycloalkanes.

1. 1,2-dichlorcyclobutane

2. 1-bromo-4-methylcyclohexane

3. 1,3-dichlorocyclopentane

4. 1-bromo-3-chloro-5-methylcyclohexane

Answers

1. (1,2-dichlorocyclobutane structure: cyclobutane with Cl on adjacent carbons)

Cl
Cl

2. Br / CH_3

3. Cl / Cl

4. Br / H_3C / Cl

◆ **Learning Exercise 12.5C**

Write the constitutional isomers and names for the cycloalkanes of C_4H_8.

Answers

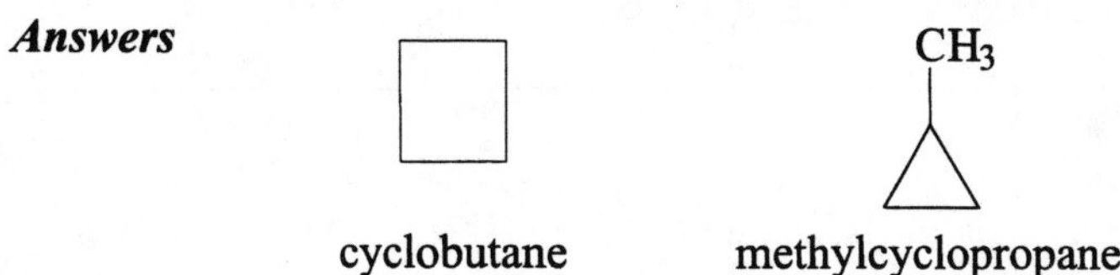

◆ **Learning Exercise 12.5D**

Name each of the following stereoisomers as the cis or trans isomer.

a. Cl Cl

b. CH_3 CH_3

c. Br Br

d. Br Cl

Answers

a. *cis*-1,2-dichlorocyclobutane	**b.** *trans*-1,3-dimethylcyclopentane
c. *cis*-1,2-dibromocyclohexane	**d.** *trans*-1-bromo-3-chlorocyclopentane

12.6 Physical Properties of Alkanes and Cycloalkanes

- The alkanes are nonpolar, less dense than water, and mostly unreactive, except that they burn vigorously.
- Alkanes are found in natural gas, gasoline, and diesel fuels.

◆ Learning Exercise 12.6

For each of the following pairs of hydrocarbons, indicate the one that has the higher boiling point.

1. butane or propane ____________ **2.** butane or cyclobutane ____________

3. 3-methylpentane or hexane ____________ **4.** cyclopentane or pentane ____________

Answers **1.** butane **2.** cyclobutane **3.** hexane **4.** cyclopentane

12.7 Chemical Properties of Alkanes and Cycloalkanes

- In combustion, an alkane or cycloalkane at a high temperature reacts rapidly with oxygen to produce carbon dioxide, water, and a great amount of heat.
- Alkanes and cycloalkanes react with Cl_2 or Br_2 to produce halogenated alkanes or cycloalkanes.

Study Note

Example: Write the equation for the combustion of methane.
Solution: Write the molecular formulas for the reactants: methane (CH_4) and oxygen (O_2). Write the products CO_2 and H_2O and balance the equation.

$$CH_4(g) + O_2(g) \longrightarrow CO_2(g) + H_2O(g) + \text{Heat}$$
$$CH_4(g) + 2O_2(g) \longrightarrow CO_2(g) + 2H_2O(g) + \text{Heat (balanced)}$$

◆ Learning Exercise 12.7A

Write a balanced equation for the complete combustion of the following:

1. propane ____

2. hexane ____

3. pentane ____

4. cyclobutane____

Answers **1.** $C_3H_8 + 5O_2 \longrightarrow 3CO_2 + 4H_2O + \text{Heat}$ **2.** $2C_6H_{14} + 19O_2 \longrightarrow 12CO_2 + 14H_2O + \text{Heat}$
3. $C_5H_{12} + 8O_2 \longrightarrow 5CO_2 + 6H_2O + \text{Heat}$ **4.** $C_4H_8 + 6O_2 \longrightarrow 4CO_2 + 4H_2O + \text{Heat}$

◆ Learning Exercise 12.7B

Write the condensed structural formulas for the monobrominated product of each of the following in the presence of light or heat:

1. ethane **2.** cyclohexane

3 methane **4.** cyclopentane

Answers

1. CH_3-CH_2-Br 2. 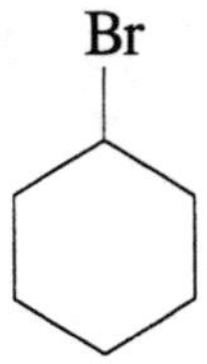 (cyclohexane ring with Br) 3. CH_3-Br 4. 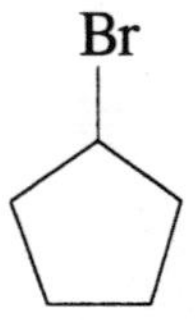 (cyclopentane ring with Br)

Check List for Chapter 12

You are ready to take the practice test for Chapter 12. Be sure that you have accomplished the following learning goals for this chapter. If you are not sure, review the section listed at the end of the goal. Then apply your new skills and understanding to the practice test.

After studying Chapter 12, I can successfully:

____ Draw the complete structural formula and the condensed structural formula for an alkane (12.1).

____ Use the IUPAC system to write the names for branched and unbranched alkanes (12.2).

____ Draw the structural formulas of alkanes from the name. (12.3).

____ Name or draw the structural formula of a haloalkane (12.4).

____ Name or draw the structural formula of a cycloalkane (12.5).

____ Describe physical properties of alkanes and cycloalkanes (12.6).

____ Write equations for the halogenation and combustion of alkanes and cycloalkanes (12.7).

Practice Test for Chapter 12

Match the name of the hydrocarbon with each of the following structures.

A. methane **B.** ethane **C.** propane **D.** pentane **E.** heptane

1. ____ $CH_3-CH_2-CH_3$

2. ____ $CH_3-CH_2-CH_2-CH_2-CH_2-CH_2-CH_3$

3. ____ CH_4

4. ____ $CH_3-CH_2-CH_2-CH_2-CH_3$

5. ____ CH_3-CH_3

Match the name of the hydrocarbons with each of the following structures.

A. butane **B.** methylcyclohexane **C.** cyclopropane
D. 3,5-dimethylhexane **E.** 2,4-dimethylhexane

6. ____ $CH_3-CH_2-CH_2-CH_3$

7. ____ $CH_3-\underset{|}{\overset{CH_3}{CH}}-CH_2-\overset{CH_3}{\underset{|}{CH}}-CH_2-CH_3$

8. ____ (cyclopropane triangle)

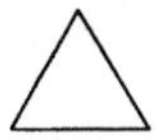

9. ____ (cyclohexane ring with CH_3)

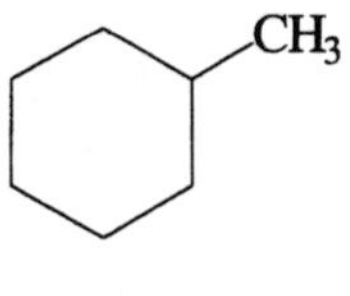

Match the name of the hydrocarbon with each of the following structures.

A. methylcyclopentane **B.** 1,2-dimethylcyclohexane **C.** cyclobutane
D. 1,1-dimethylcyclohexane **E.** ethylcyclopentane

10. ___ 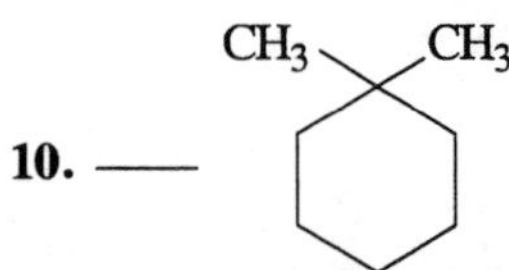

11. ___ 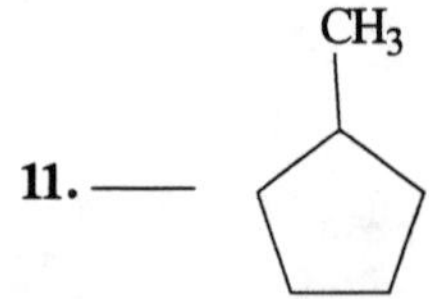

12. ___ CH_2-CH_3

Match each of the following compounds with the correct name.

A. 2,4-dichloropentane **B.** 1,3-dichlorocyclopentane **C.** 1,2-dichloropentane
D. 2,3-dichlorocyclopentane **E.** 4,5-dichloropentane

13. ___ $CH_3-CH_2-CH_2-CH(Cl)-CH_2-Cl$

14. ___ 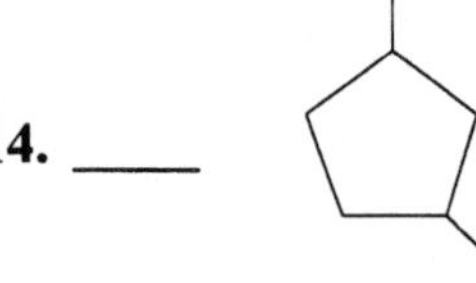

15. ___ 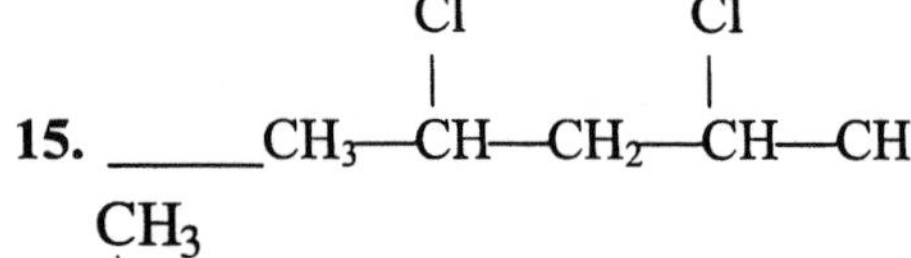

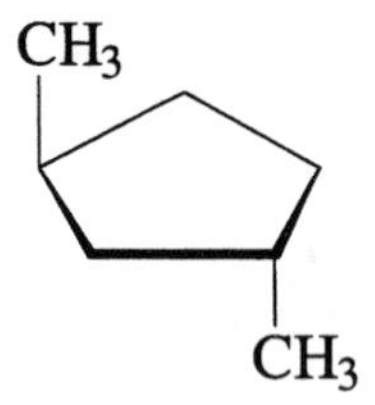

16. The above compound is named

A. 1,3-dimethylcyclopentane.
B. *cis*-1,3-dimethylcyclopentane.
C. *cis*-1,4-dimethylcyclopentane.
D. *trans*-1,3-dimethylcyclopentane.
E. *trans*-1,4-dimethylcyclopentane.

17.

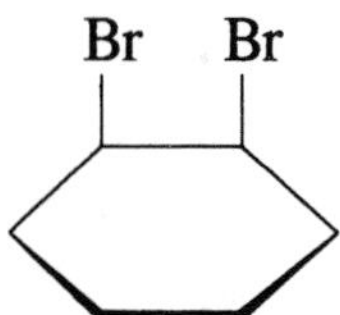

The above compound is named

A. 1,3-dibromocyclohexane.
B. *cis*-1,2-dibromocyclohexane.
C. *cis*-1,3-dibromocyclohexane.
D. *trans*-1,1-dibromocyclohexane.
E. *trans*-1,2-dibromocyclohexane.

18. The correctly balanced equation for the complete combustion of ethane is

A. $C_2H_6 + O_2 \longrightarrow 2CO + 3H_2O$.
B. $C_2H_6 + O_2 \longrightarrow CO_2 + H_2O$.
C. $C_2H_6 + 2O_2 \longrightarrow 2CO_2 + 3H_2O$.
D. $2C_2H_6 + 7O_2 \longrightarrow 4CO_2 + 6H_2O$.
E. $2C_2H_6 + 4O_2 \longrightarrow 4CO_2 + 6H_2O$.

19. The hydrocarbon with the highest boiling point is

A. pentane. **B.** cyclopentane. **C.** cyclohexane.
D. 2-Methylbutane. **E.** hexane.

Identify the relationship of each of the following pairs as

A. constitutional isomers.
B. the same compound.
C. cis-trans isomers.
D. different compounds.

20. _____ 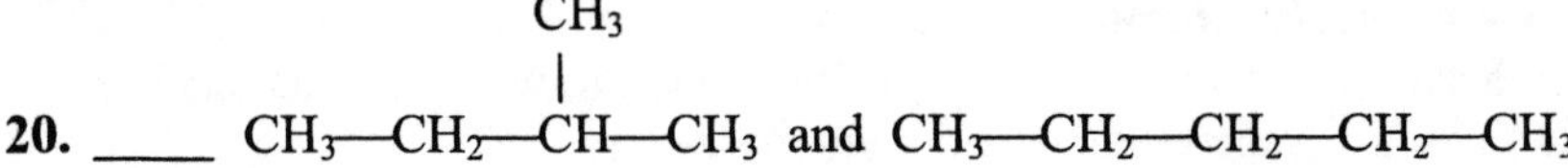

21. _____ 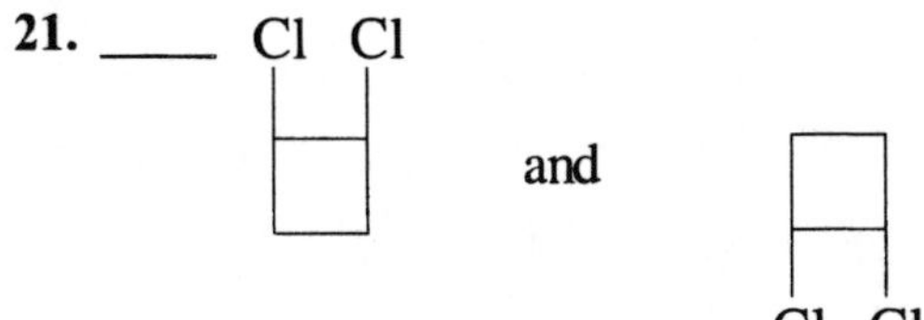

22. _____ 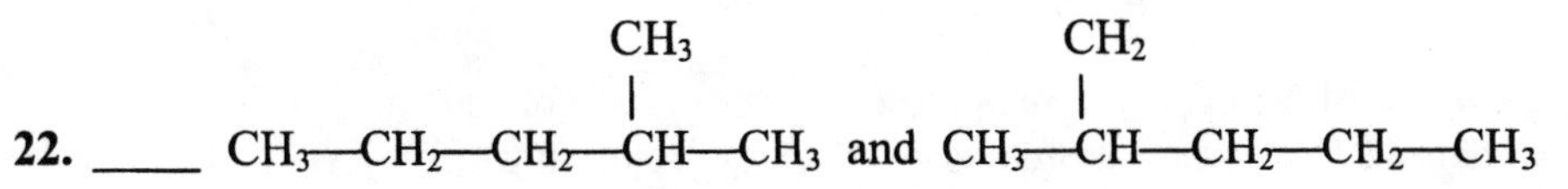

23. _____ $CH_3—CH_2—CH_2—CH_2—CH_2—CH_3$ and $CH_3—CH_2—CH_2—CH_2—CH_3$

24. _____ 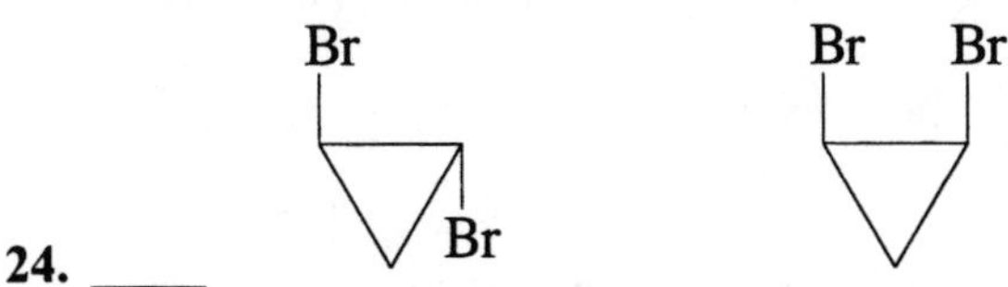

25. The number of hydrogen atoms in an alkane with 7 carbon atoms is

A. 7 **B.** 10 **C.** 14 **D.** 16 **E.** 20

Answers to the Practice Test

1. C	**2.** E	**3.** A	**4.** D	**5.** B
6. A	**7.** E	**8.** C	**9.** B	**10.** D
11. A	**12.** E	**13.** C	**14.** B	**15.** A
16. D	**17.** B	**18.** D	**19.** C	**20.** A
21. B	**22.** B	**23.** D	**24.** C	**25.** D

Answers and Solutions to Selected Text Problems

12.1 **a.** The formula C_nH_{2n+2} indicates an alkane with a formula C_7H_{16}
b. The formula C_nH_{2n+2} indicates $2(5) + 2 = 12$ H atoms
c. The formula C_nH_{2n+2} indicates H atoms $= (10 - 2)/2 = 4$ C atoms

12.3 **a.** In expanded formulas, each C—H and each C—C bond is drawn separately.

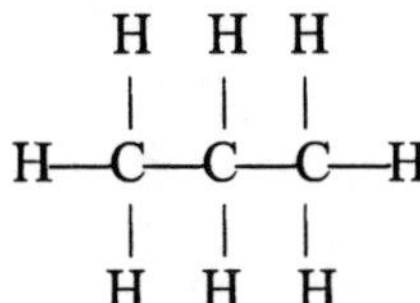

b. A condensed formula groups hydrogen atoms with each carbon atom.
$CH_3—CH_2—CH_2—CH_2—CH_2—CH_3$

c. A line-bond formula shows only the bonds connecting the carbon atoms.

12.5 **a.** Pentane is a carbon chain of five (5) carbon atoms.
b. Heptane is a carbon chain of seven (7) carbon atoms.
c. Hexane is a carbon chain of six (6) carbon atoms.

12.7 Constitutional isomers have the same molecular formula, but different structural formulas. A different conformation of the same structure occurs when rotation about a single bond moves the attached groups into different positions, but does not change the arrangement of atoms.

a. The same structure with two different conformations
b. Same molecular formula, but different structures, which are constitutional isomers.
c. Same molecular formula, but different structures, which are constitutional isomers.

12.9 The alkyl name for a hydrocarbon substituent uses the name of the alkane, but changes the ending to –yl.

a. propyl (from propane) **b.** isopropyl
c. butyl (from butane) **d.** methyl (from methane)

12.11 **a.** 2-methylpropane has a chain of three carbon atoms with a methyl group on carbon 2.
b. 2-methylpentane has a chain of five carbon atoms with a methyl group on carbon 2.
c. 4-ethyl-2-methylhexane has a chain of six carbon atoms with a methyl group on carbon 2 closest to the end of the main chain, and an ethyl group on carbon 4. Ethyl is listed first alphabetically.
d. 2,3,4-trimethylheptane is a main chain of seven carbon atoms with methyl groups on carbons 2, 3, and 4.
e. 5-isopropyl-3-methyloctane is a main chain of eight carbons with a methyl group on carbon 3 and an isopropyl group (named first alphabetically) on carbon 5.

12.13 For line-bond formulas, carbon atoms are counted from the end of the chain and at each corner.
a. 2,3-dimethylpentane **b.** 3-ethyl-5-methylheptane
c. 3-isopropyl-2,4-dimethylpentane

12.15 Draw the main chain with the number of carbon atoms in the ending. For example, butane has a main chain of 4 carbon atoms, and hexane has a main chain of 6 carbon atoms. Attach substituents on the carbon atoms indicated. For example, in 3-methylpentane, a CH_3— group is bonded to carbon 3 of a five-carbon chain.

a.
$$\begin{array}{l} CH_3 \\ | \\ CH_3-CH-CH_2-CH_3 \end{array}$$

b.
$$\begin{array}{l} CH_3 \\ | \\ CH_3-CH_2-C-CH_2-CH_3 \\ | \\ CH_3 \end{array}$$

c.
$$\begin{array}{l} CH_3\ CH_3\qquad\quad CH_3 \\ |\qquad|\qquad\qquad\ | \\ CH_3-CH-CH-CH_2-CH-CH_3 \end{array}$$

d.
$$\begin{array}{l} CH_3\ CH_2-CH_3\ CH_3 \\ |\qquad|\qquad\qquad\ | \\ CH_3-CH-CH-CH_2-CH-CH_2-CH_2-CH_3 \end{array}$$

e.
$$\begin{array}{l} CH_3 \\ | \\ CH_3\qquad\qquad\ CH-CH_3 \\ |\qquad\qquad\quad\ | \\ CH_3-CH-CH_2-CH-CH_2-CH_2-CH_3 \end{array}$$

f.
$$\begin{array}{l} CH_3 \\ | \\ CH_2-CH_2-CH_3 \\ | \\ CH_3-CH_2-CH_2-CH-CH_2-CH_2-CH_2-CH_2-CH_3 \end{array}$$

12.17 **a.** This problem wants the names of all the constitutional isomers that have a methyl group bonded to a seven-carbon chain: 2-methylheptane; 3-methylheptane; 4-methylheptane

b. This problem wants the names of all the constitutional isomers that have two methyl groups or an ethyl group bonded to a five-carbon chain; 2,2-dimethylpentane; 3,3-dimethylpentane; 2,3-dimethylpentane; 2,4-dimethylpentane; 3-ethylpentane.

12.19 In the IUPAC system, the halogen substituent in haloalkanes is named as a halo- on the main chain of carbon atom; simple haloalkanes use common alkyl names followed by the name of the halogen.

a. bromoethane, ethyl bromide
b. 1-fluoropropane, propyl fluoride
c. 2-chloropropane, isopropyl chloride
d. trichloromethane, chloroform

12.21 **a.** 2-bromo-3-methylbutane
b. 3-bromo–2–chloropentane
c. 2-fluoro-2-methylbutane

12.23 **a.**
Cl
|
$CH_3—CH—CH_3$

b.
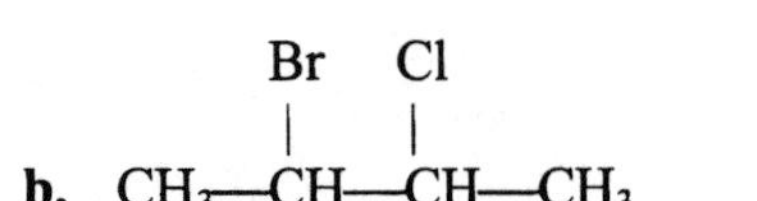
Br Cl
| |
$CH_3—CH—CH—CH_3$

c. CH_3Br

d. $CH_3—CH_2—CH_2—CH_2—Cl$

e.
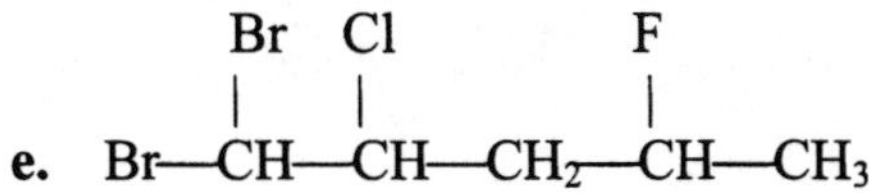
Br Cl F
| | |
$Br—CH—CH—CH_2—CH—CH_3$

f. CBr_4

12.25 Methyl chloride is CH_3Cl; ethyl chloride is $CH_3—CH_2Cl$.

12.27 The isomers are written first using a four-carbon chain of butane and then with the three-carbon chain with a methyl group. To each carbon chain, a chlorine atom is attached to give different isomers.

$Cl—CH_2—CH_2—CH_2—CH_3$
1-chlorobutane

Cl
|
$CH_3—CH—CH_2—CH_3$
2-chlorobutane

CH_3
|
$Cl—CH_2—CH—CH_3$
1-chloro-2-methylpropane

Cl
|
$CH_2—C—CH_3$
|
CH_3
2-chloro-2-methylpropane

12.29 The general formula for cycloalkanes is C_nH_{2n}.

a. $n = 5$ C atoms; $2n = 10$ H atoms; C_5H_{10}
b. $n = 4$ C atoms; $2n = 8$ H atoms; C_4H_8
c. $2n = 12$ H atoms; $n = 6$ C atoms; C_6H_{12}

12.31 **a.** A ring of four carbon atoms is cyclobutane.
b. A ring of five carbon atoms with one chlorine atom is chlorocyclopentane; no numbering is needed for a single substituent.
c. A ring of six carbon atoms with one methyl group is methylcyclohexane; no numbering is needed for a single substituent.
d. 1-bromo-3-methylcyclobutane
e. 1-bromo-2-chlorocyclopentane
f. 1,3-dibromo-5-methylcyclohexane

12.33 Draw the cyclic structure first, and then attach the substituents. When there are two or more substituents, start with the first on a carbon assigned number 1 and continue around the ring.

a.

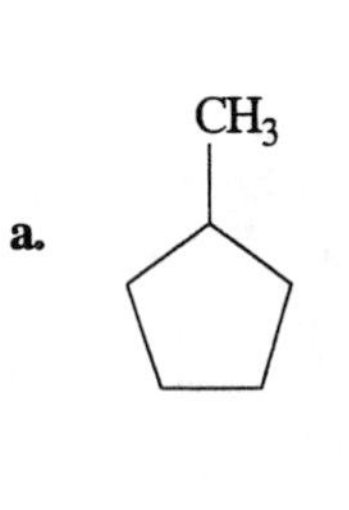

b.

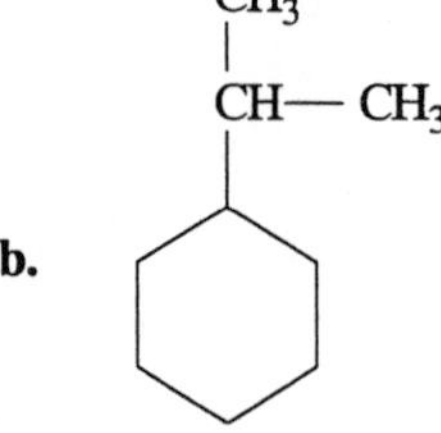

c.

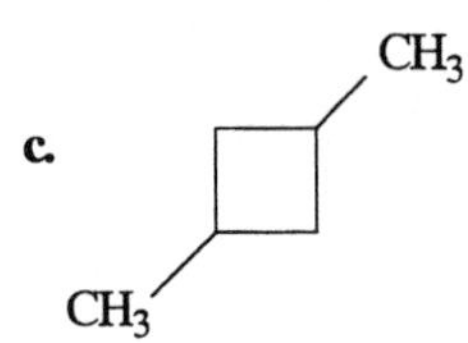

d.

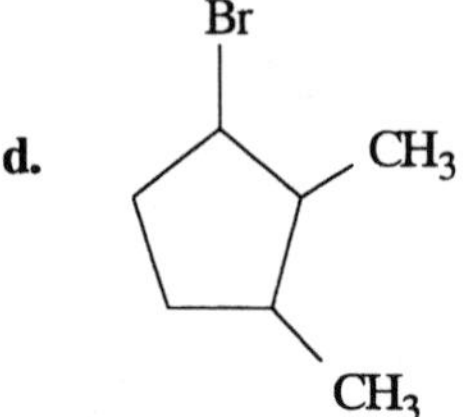

12.35 Four constitutional isomers are possible: 1,1-dimethylcyclopropane, 1,2-dimethylcyclopropane, ethylcyclopropane, methylcyclobutane.

12.37 Cis and trans isomers differ in the orientation of atoms in space. In cis isomer, two atoms are on the same side; in the trans isomer, two atoms are on opposite sides (up and down).

a. Does not have *cis-,trans* isomers
b. *trans*-1,2-dimethylcyclopropane
c. *cis*-1,2-dichlorocyclobutane
d. *trans*-1,3-dimethylcyclopentane

12.39 In a cis isomer, two atoms or groups are written on the same side (both up or both down). In the trans isomer, two atoms or groups are written on opposite sides (up and down).

a.

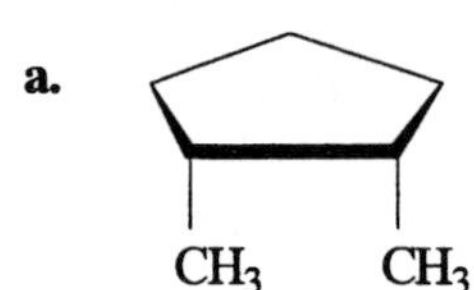

b.

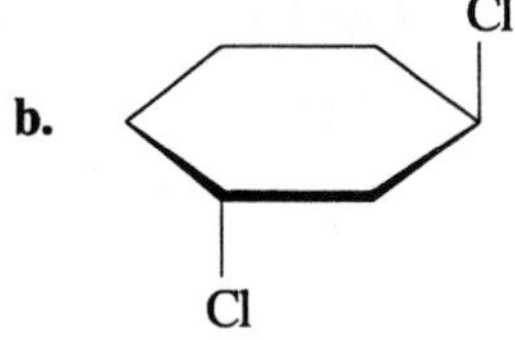

c.

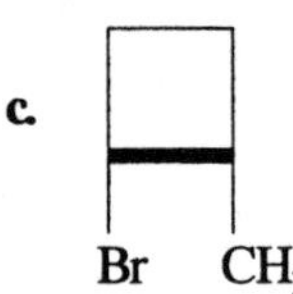

12.41 Alkanes are nonpolar and less dense than water.

a. $CH_3—CH_2—CH_2—CH_2—CH_2—CH_2—CH_3$
b. liquid
c. Insoluble in water
d. Float
e. Lower

12.43 Longer carbon chains have higher boiling points. The boiling points of branched alkanes are usually lower than the same number of carbon atoms in a continuous chain. Cycloalkanes have higher boiling points than continuous-chain alkanes.

a. Heptane has a longer carbon chain.
b. Cyclopropane
c. The continuous chain hexane has a higher boiling point than its branched chain isomer.

12.45 In combustion, a hydrocarbon reacts with oxygen to yield CO_2 and H_2O.

a. $2C_2H_6 + 7O_2 \longrightarrow 4CO_2 + 6H_2O$
b. $2C_3H_6 + 9O_2 \longrightarrow 6CO_2 + 6H_2O$
c. $2C_8H_{18} + 25O_2 \longrightarrow 16CO_2 + 18H_2O$
d. $C_6H_{12} + 9O_2 \longrightarrow 6CO_2 + 6H_2O$

12.47 **a.** $CH_3—CH_2—Cl$ **b.** Cl

c.

$$\begin{array}{c} CH_3 \\ | \\ CH_3—CH—CH_2—Cl \end{array} \qquad \begin{array}{c} CH_3 \\ | \\ CH_3—C—CH_3 \\ | \\ Cl \end{array}$$

12.49 Constitutional isomers have the same molecular formulas, but different arrangements of atoms.

a. constitutional isomers **b.** constitutional isomers
c. same molecule **d.** constitutional isomers

12.51 The alkyl name for a hydrocarbon substituent uses the name of the alkane, but changes the ending to –yl.

a. methyl **b.** propyl **c.** isopropyl

12.53 Identify the longest carbon chain and number it from the end closest to the first substituent. Use the prefix di- when two substituents are identical.

a. 2,2-dimethylbutane **b.** chloroethane
c. 2-bromo-4-ethylhexane **d.** 1,1-dibromocyclohexane

12.55 Write the carbon atom in the main chain first. Attach the substituents listed in front of the alkane name or use the alkyl group indicated.

a.

$$\begin{array}{c} CH_2—CH_3 \\ | \\ CH_3—CH_2—CH—CH_2—CH_2—CH_3 \end{array}$$

b.

$$\begin{array}{c} CH_3 \quad CH_3 \\ | \qquad | \\ CH_3—CH—CH—CH_2—CH_3 \end{array}$$

c.

$$\begin{array}{c} Cl \\ | \\ Cl—CH_2—CH_2—C—CH_2—CH_2—CH_2—CH_3 \\ | \\ CH_3 \end{array}$$

d.

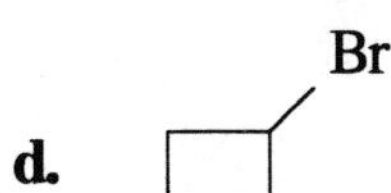

e.

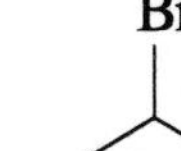

Br

$$\begin{array}{l} CH—CH_3 \\ | \\ CH_3 \end{array}$$

f.

$$\begin{array}{c} CH_3 \\ | \\ CH_3—CH_2—CH—Cl \end{array}$$

12.57 A line-bond formula shows only the bonds connecting the carbon atoms. The number of bonds to hydrogen atoms is understood.

a. **b.** **c.** Br

12.59 **a.** You should have two of the following four possibilities.

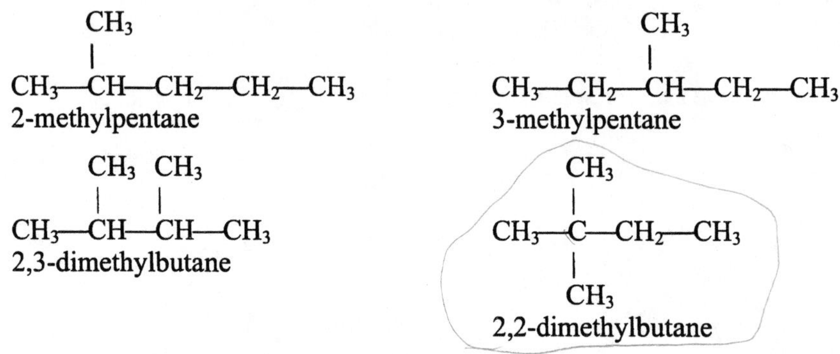

b. The molecular formula C_7H_{16} has the following constitutional isomers:

$CH_3—CH_2—CH_2—CH_2—CH_2—CH_2—CH_3$ heptane

CH_3
$CH_3—CH—CH_2—CH_2—CH_2—CH_3$ 2-methylhexane

CH_3
$CH_3—CH_2—CH—CH_2—CH_2—CH_3$ 3-methylhexane

CH_3
$CH_3—C—CH_2—CH_2—CH_3$ 2,2-dimethylpentane
CH_3

CH_3
$CH_3—CH_2—C—CH_2—CH_3$ 3,3-dimethylpentane
CH_3

CH_3 CH_3
$CH_3—CH—CH—CH_2—CH_3$ 2,3-dimethylpentane

CH_3 CH_3
$CH_3—CH—CH_2—CH—CH_3$ 2,4-dimethylpentane

CH_3 CH_3
$CH_3—C—CH—CH_3$ 2,2,3-trimethylbutane
CH_3

c. $CH_3—CH_2—CH_2—CH_2—CH_2—CH_3$ hexane

CH_3
|
$CH_3—CH—CH_2—CH_2—CH_3$ 2-methylpentane

CH_3
|
$CH_3—CH_2—CH—CH_2—CH_3$ 3-methylpentane

CH_3 CH_3
| |
$CH_3—CH—CH—CH_3$ 2,3-dimethylbutane

d.

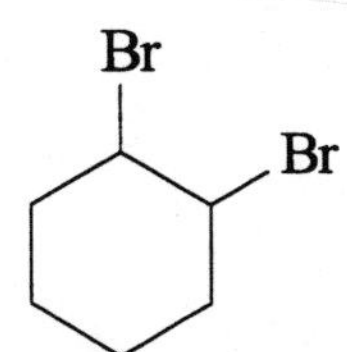

1,2-dibromocyclohexane

Br

Br

1,3-dibromocyclohexane

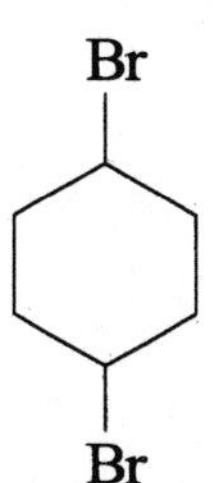

1,4-dibromocyclohexane

12.61 Draw the structure and then write the correct name using IUPAC rules.

a. 2-methylbutane
b. 2,3-dimethylpentane
c. 1,3-dibromocyclohexane
d. hexane

12.63 All of the following are constitutional isomers of $C_5H_{11}Cl$. You should have six.

$CH_3—CH_2—CH_2—CH_2—CH_2—Cl$

Cl
|
$CH_3—CH—CH_2—CH_2—CH_3$

Cl
|
$CH_3—CH_2—CH—CH_2—CH_3$

CH_3
|
$CH_3—CH—CH_2—CH_2—Cl$

CH_3 Cl
| |
$CH_3—CH—CH—CH_3$

CH_3
|
$Cl—CH_2—CH—CH_2—CH_3$

CH_3
|
$CH_3—C—CH_2—CH_3$
|
Cl

CH_3
|
$CH_3—C—CH_2—Cl$
|
CH_3

12.65 **a.** 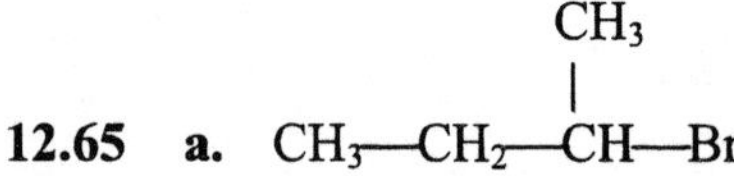

b. 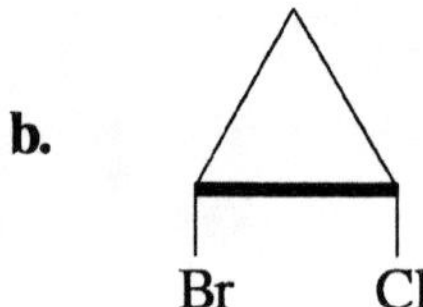

c. 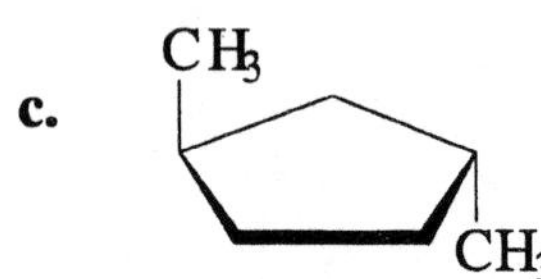

d.

Br Br

12.67 Condensed structural formula $CH_3-C(CH_3)_2-CH_2-CH(CH_3)-CH_3$

molecular formula of C_8H_{18}

The combustion reaction: $2C_8H_{18} + 25O_2 \rightarrow 16CO_2 + 18H_2O$

12.69 **a.** heptane **b.** cyclopentane
c. hexane **d.** cyclohexane

12.71 **a.** $C_3H_8 + 5O_2 \longrightarrow 3CO_2 + 4H_2O$ **b.** $C_5H_{12} + 8O_2 \longrightarrow 5CO_2 + 6H_2O$
c. $C_4H_8 + 6O_2 \longrightarrow 4CO_2 + 4H_2O$ **d.** $2C_8H_{18} + 25O_2 \longrightarrow 16CO_2 + 18H_2O$

12.73 **a.** CH_3-CH_2-Cl

b. $CH_3-CH_2-CH_2-Cl$ and $CH_3-CH(Cl)-CH_3$

c. 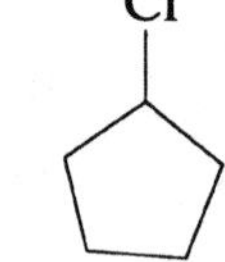

12.75 **a.** $C_5H_{12} + 8O_2 \longrightarrow 5CO_2 + 6H_2O$
b. 72.0 g/mole

c. $1\ \text{gal} \times \frac{3.78\ \text{L}}{1\ \text{gal}} \times \frac{1000\ \text{mL}}{1\ \text{L}} \times \frac{0.63\ \text{g}}{1\ \text{mL}} \times \frac{1\ \text{mole } C_5H_{12}}{72.0\ \text{g}} \times \frac{845\ \text{kcal}}{1\ \text{mole}} = 2.8 \times 10^4\ \text{kcal}$

d. $1\ \text{gal} \times \frac{3.78\ \text{L}}{1\ \text{gal}} \times \frac{1000\ \text{mL}}{1\ \text{L}} \times \frac{0.63\ \text{g}}{1\ \text{mL}} \times \frac{1\ \text{mole } C_5H_{12}}{72.0\ \text{g}} \times \frac{5\ \text{moles } CO_2}{1\ \text{mole } C_5H_{12}} \times \frac{22.4\ \text{L}}{1\ \text{mole } CO_2} = 3700\ \text{L}$

13
Unsaturated Hydrocarbons

Study Goals

- Classify unsaturated compounds as alkenes, cycloalkenes, and alkynes.
- Write IUPAC and common names for alkenes and alkynes.
- Write structural formulas and names for cis-trans isomers of alkenes.
- Write equations for halogenation, hydration, and hydrogenation of alkenes and alkynes.
- Describe the formation of a polymer from alkene monomers.
- Describe the bonding in benzene.
- Write structural formulas and give the names of aromatic compounds.
- Describe the physical and chemical properties of aromatic compounds.

Think About It

1. The label on a bottle of vegetable oil says the oil is unsaturated. What does this mean?

2. What are polymers?

3. A margarine is partially hydrogenated. What does that mean?

Key Terms

Match the statements shown below with the following key terms.

a. alkene **b.** hydrogenation **c.** alkyne **d.** hydration **e.** polymer

1. _____ A long-chain molecule formed by linking many small molecules

2. _____ The addition of H_2 to a carbon-carbon double bond

3. _____ An unsaturated hydrocarbon containing a carbon-carbon double bond

4. _____ The addition of H_2O to a carbon-carbon double bond

5. _____ A compound that contains a triple bond

Answers **1.** e **2.** b **3.** a **4.** d **5.** c

13.1 Alkenes and Alkynes

- Alkenes are unsaturated hydrocarbons that contain one or more carbon–carbon double bonds.
- In alkenes, the three groups bonded to the carbons in the double bond are planar and arranged at angles of 120°.
- Alkynes are unsaturated hydrocarbons that contain a carbon–carbon triple bond.
- The atoms bonded to a carbon–carbon triple bond are linear.

◆ Learning Exercise 13.1

Classify the following structural formulas as alkane, alkene, cycloalkene, or alkyne

1. ____ $CH_3—CH_2—CH_3$

2. ____

3. ____ $CH_3—C{\equiv}C—CH_3$

4. ____ $CH_3—CH_2—CH{=}\overset{\displaystyle CH_3 \atop |}{C}—CH_2—CH_3$

Answers **1.** alkane **2.** alkene **3.** alkyne **4.** alkene

13.2 Naming Alkenes and Alkynes

- The IUPAC names of alkenes are derived by changing the *ane* ending of the parent alkane to *ene*.
- For example, the IUPAC name of $H_2C{=}CH_2$ is ethene. It has a common name of ethylene.
- In alkenes, the longest carbon chain containing the double bond is numbered from the end nearest the double bond. In cycloalkenes with substituents, the double bond carbons are given positions of 1 and 2, and the ring numbered to give the next lower numbers to the substituents.

$CH_3—CH{=}CH_2$ Propene (propylene)

$CH_2{=}CH—CH_2—CH_3$ 1-Butene

$CH_3—CH{=}\overset{\displaystyle CH_3 \atop |}{C}—CH_3$ 2-Methylbutene

- The alkynes are a family of unsaturated hydrocarbons that contain a triple bond. They use naming rules similar to the alkenes, but the parent chain ends with *yne.*

$HC{\equiv}CH$ ethyne

$CH_3—C{\equiv}CH$ propyne

◆ Learning Exercise 13.2A

Write the IUPAC (and common name, if one) for each of the following alkenes:

1. $CH_3CH{=}CH_2$

2. $CH_3—CH{=}CH—CH_3$

3. 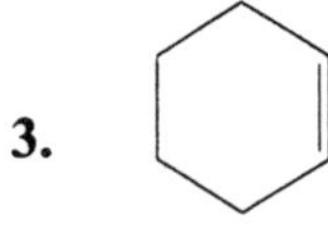

4. $CH_2{=}CH\overset{\displaystyle Cl \atop |}{C}HCH_2\overset{\displaystyle CH_3 \atop |}{C}HCH_3$

5. $CH_3—CH{=}\overset{\displaystyle CH_3 \atop |}{C}—CH_2—CH_3$

6. $CH_3—CH_2—\overset{\displaystyle CH_2 \atop \|}{C}H$

Answers **1.** propene (propylene) **2.** 2-butene **3.** cyclohexene
4. 3-chloro-5-methyl-1-hexene **5.** 3-methyl-2-pentene **6.** 1-butene

◆ Learning Exercise 13.2B

Write the IUPAC and common name (if any) of each of the following alkynes.

1. $HC{\equiv}CH$
2. $CH_3{-}C{\equiv}CH$
3. $CH_3{-}CH_2{-}C{\equiv}CH$
4. $CH_3{-}\underset{}{\overset{CH_3}{\overset{|}{C}H}}{-}C{\equiv}C{-}CH_3$

Answers

1. ethyne (acetylene)
2. propyne (methylacetylene)
3. 1-butyne (ethylacetylene)
4. 4-methyl-2-pentyne

◆ Learning Exercise 13.2C

Draw the condensed structural formula for each of the following:

1. 2-pentyne
2. 2-chloro-2-butene
3. 3-bromo-2-methyl-2-pentene
4. 3-methylcyclohexene

Answers

1. $CH_3{-}C{\equiv}C{-}CH_2{-}CH_3$
2. $CH_3{-}CH{=}\overset{Cl}{\overset{|}{C}}{-}CH_3$
3. $CH_3{-}\underset{\underset{CH_3}{|}}{C}{=}\overset{Br}{\overset{|}{C}}{-}CH_2{-}CH_3$
4. (cyclohexene ring with CH_3 at position 3)

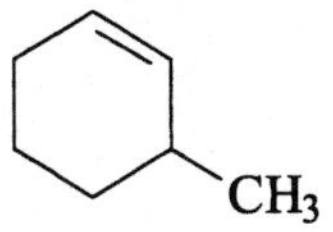

13.3 Cis–Trans Isomers

- Cis–trans isomers are possible for alkenes because there is no rotation around the rigid double bond.
- In the cis isomer, groups are attached on the same side of the double bond, whereas in the trans isomer, they are attached on the opposite sides of the double bond.

◆ Learning Exercise 13.3A

Write the cis–trans isomers of 2,3-dibromo-2-butene and name each.

Answers

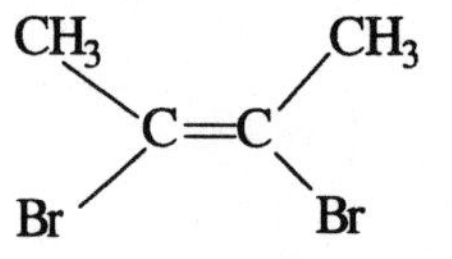

```
CH3        CH3          CH3        Br
   \      /                \      /
    C=C                     C=C
   /      \                /      \
Br         Br            Br        CH3
```

cis-2,3-dibromo-2-butene 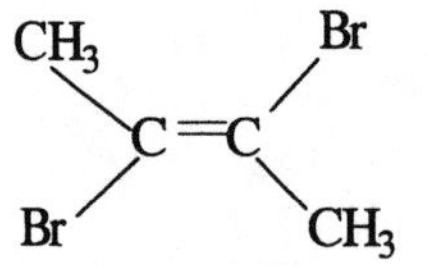　　　*trans*-2,3-dibromo-2-butene

In the cis isomer, the bromine atoms are attached on the same side of the double bond; in the trans isomer, they are on opposite sides.

◆ Learning Exercise 13.3B

Name the following alkenes using the cis–trans where isomers are possible.

```
1.  Br    Br            2.  H     CH3
     \   /                   \   /
      C-C                     C=C
     /   \                   /   \
    H     H               H3C     H

3.  Cl    H             4.  H     CH2—CH3
     \   /                   \   /
      C=C                     C=C
     /   \                   /   \
  H3C     H               H3C     H
```

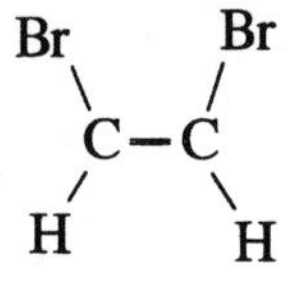

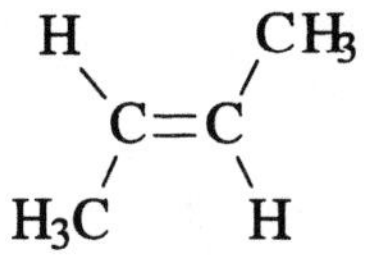

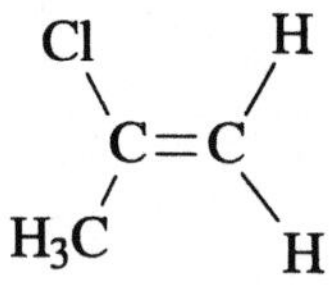

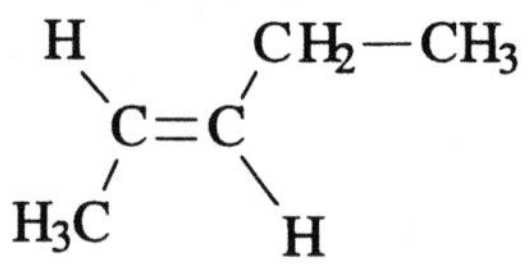

Answers

1. *cis*-1,2-dibromoethene　　**2.** *trans*-2-butene
3. 2-chloropropene (not a cis–trans isomer)　　**4.** *trans*-2-pentene

13.4 Addition Reactions

- The addition of small molecules to the double bond is a characteristic reaction of alkenes.
- Hydrogenation adds hydrogen atoms to the double bond of an alkene or the triple bond of an alkyne to yield an alkane.

$$CH_2{=}CH_2 + H_2 \xrightarrow{Pt} CH_3{-}CH_3$$

$$HC{\equiv}CH + 2H_2 \longrightarrow CH_3{-}CH_3$$

- Halogenation adds bromine or chlorine atoms to produce dihaloalkanes.

$$CH_2{=}CH_2 + Br_2 \longrightarrow Br{-}CH_2{-}CH_2{-}Br$$

- Hydrohalogenation adds hydrogen halides and hydration adds water to a double bond.

$$CH_2{=}CH_2 + HCl \longrightarrow CH_3{-}CH_2{-}Cl$$

$$CH_2{=}CH_2 + HOH \xrightarrow{H^+} CH_3{-}CH_2{-}OH$$

- According to Markovnikov's rule, the H from the reactant (HX or HOH) bonds to the carbon in the double bond that has the greater number of hydrogen atoms.

◆ Learning Exercise 13.4A

Write the products of the following addition reactions.

1. $CH_3{-}CH_2{-}CH{=}CH_2 + H_2 \xrightarrow{Pt}$

2. 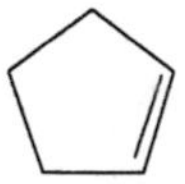 $+ \; H_2 \xrightarrow{Pt}$

3. $CH_3CH{=}CHCH_2CH_3 \; + \; Cl_2 \longrightarrow$

4. $CH_3CH{=}CH_2 \; + \; H_2 \xrightarrow{Pt}$

5. $CH_3CH{=}CHCH_3 \; + \; Br_2 \longrightarrow$

Answers

1. $CH_3{-}CH_2{-}CH_2{-}CH_3$
2. 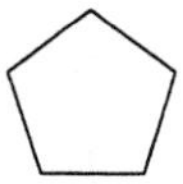
3. $CH_3{-}CH(Cl){-}CH(Cl){-}CH_2{-}CH_3$
4. $CH_3{-}CH_2{-}CH_3$
5. $CH_3{-}CH(Br){-}CH(Br){-}CH_3$

◆ **Learning Exercise 13.4B**

1. $CH_3—CH=CH—CH_3 + HCl \longrightarrow$

2. 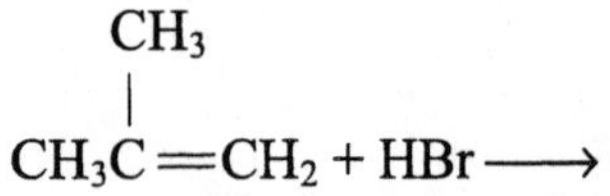 $CH_3\overset{\overset{CH_3}{|}}{C}=CH_2 + HBr \longrightarrow$

3. $CH_3CH=CH_2 + HOH \xrightarrow{H^+}$

4. 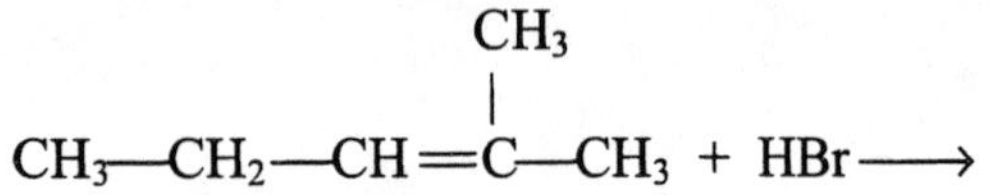 $CH_3—CH_2—CH=\overset{\overset{CH_3}{|}}{C}—CH_3 + HBr \longrightarrow$

5. 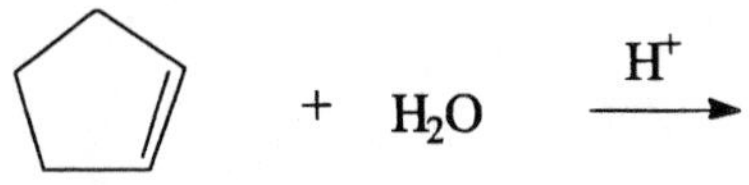 (cyclopentene) $+ H_2O \xrightarrow{H^+}$

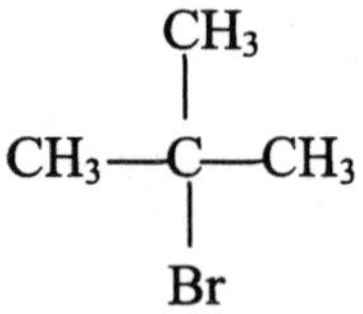

Answers

1. 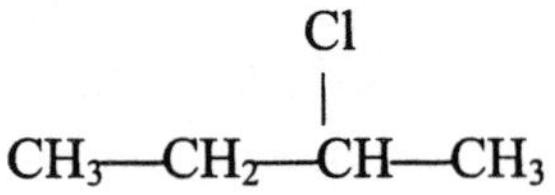 $CH_3—CH_2—\overset{\overset{Cl}{|}}{CH}—CH_3$
2. $CH_3—\underset{\underset{Br}{|}}{\overset{\overset{CH_3}{|}}{C}}—CH_3$
3. $CH_3—\overset{\overset{OH}{|}}{CH}—CH_3$ 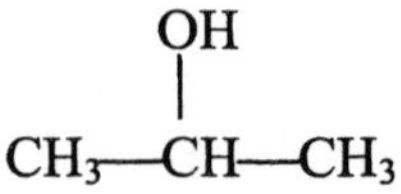
4. 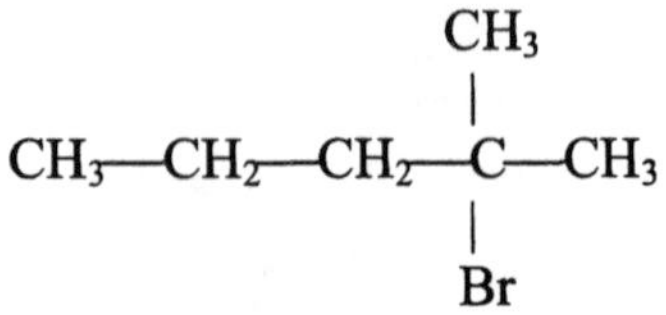 $CH_3—CH_2—CH_2—\underset{\underset{Br}{|}}{\overset{\overset{CH_3}{|}}{C}}—CH_3$
5. 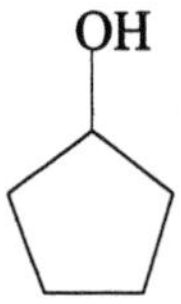 (cyclopentanol: cyclopentane ring with OH)

13.5 Polymerization

- *Polymers* are large molecules prepared from the bonding of many small units called *monomers*.
- Many synthetic polymers are made from small alkene monomers.

◆ **Learning Check 13.5A**

Write the formula of the alkene monomer that would be used for each of the following polymers.

1. 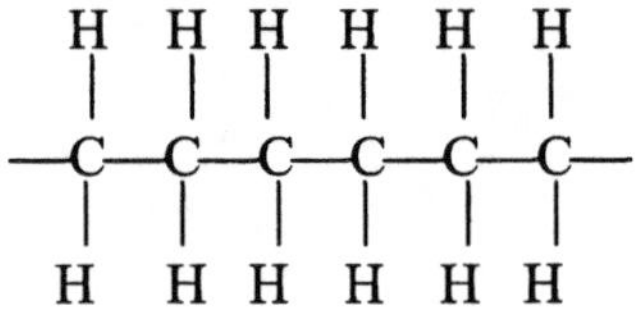 $—\underset{\underset{H}{|}}{\overset{\overset{H}{|}}{C}}—\underset{\underset{H}{|}}{\overset{\overset{H}{|}}{C}}—\underset{\underset{H}{|}}{\overset{\overset{H}{|}}{C}}—\underset{\underset{H}{|}}{\overset{\overset{H}{|}}{C}}—\underset{\underset{H}{|}}{\overset{\overset{H}{|}}{C}}—\underset{\underset{H}{|}}{\overset{\overset{H}{|}}{C}}—$

2. 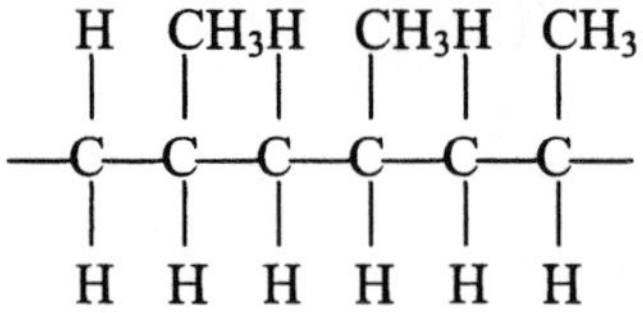 $—\underset{\underset{H}{|}}{\overset{\overset{H}{|}}{C}}—\underset{\underset{H}{|}}{\overset{\overset{CH_3}{|}}{C}}—\underset{\underset{H}{|}}{\overset{\overset{H}{|}}{C}}—\underset{\underset{H}{|}}{\overset{\overset{CH_3}{|}}{C}}—\underset{\underset{H}{|}}{\overset{\overset{H}{|}}{C}}—\underset{\underset{H}{|}}{\overset{\overset{CH_3}{|}}{C}}—$

3. $—\underset{\underset{F}{|}}{\overset{\overset{F}{|}}{C}}—\underset{\underset{F}{|}}{\overset{\overset{F}{|}}{C}}—\underset{\underset{F}{|}}{\overset{\overset{F}{|}}{C}}—\underset{\underset{F}{|}}{\overset{\overset{F}{|}}{C}}—\underset{\underset{F}{|}}{\overset{\overset{F}{|}}{C}}—\underset{\underset{F}{|}}{\overset{\overset{F}{|}}{C}}—$

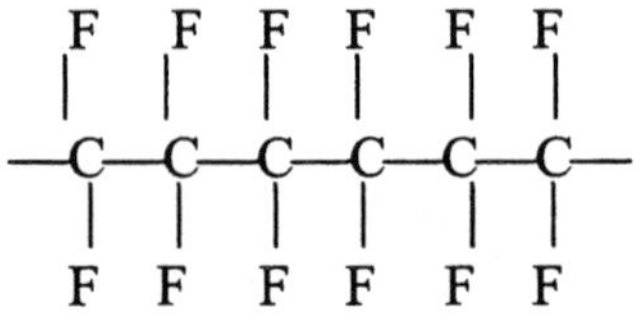

Answers 1. $H_2C=CH_2$ 2. 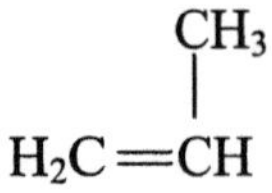 $H_2C=\overset{\overset{CH_3}{|}}{CH}$ 3. $F_2C=CF_2$

◆ Learning Check 13.5B

Write three sections of the polymer that would result when 1,1-difluoroethene is the monomer unit.

Answer

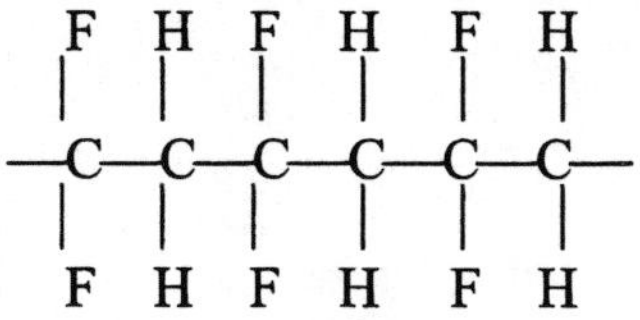

```
 F  H  F  H  F  H
 |  |  |  |  |  |
—C—C—C—C—C—C—
 |  |  |  |  |  |
 F  H  F  H  F  H
```

13.6 Aromatic Compounds

- Most aromatic compounds contain benzene, a cyclic structure containing six CH units. The structure of benzene is represented as a hexagon with a circle in the center.
- The names of many aromatic compounds use the parent name benzene, although many common names were retained as IUPAC names, such as toluene, phenol, and aniline. For two branches, the positions are often shown by the prefixes *ortho* (1,2-), *meta* (1,3-), and *para* (1,4-).

◆ Learning Exercise 13.6

Write the IUPAC (or common name) for each of the following.

1.

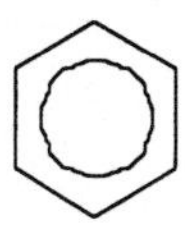

2. 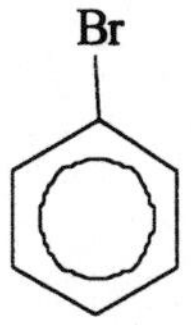 Br

3. CH_3

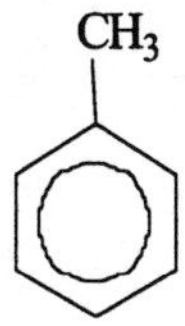

4. 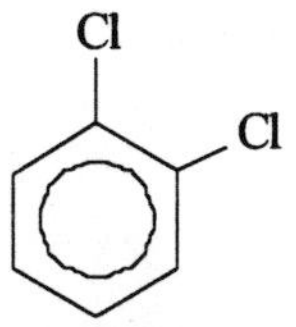 Cl Cl

5. 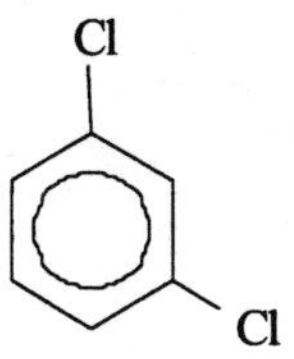 Cl Cl

6. 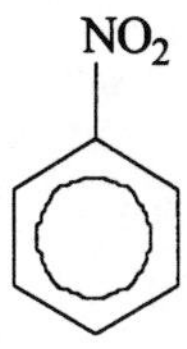 NO_2

7. 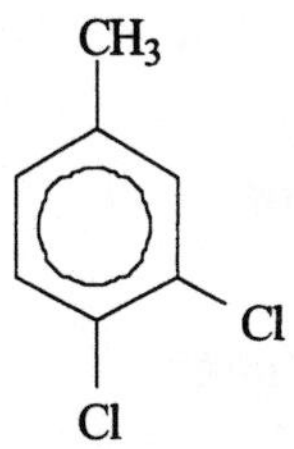 CH_3 Cl Cl

8. 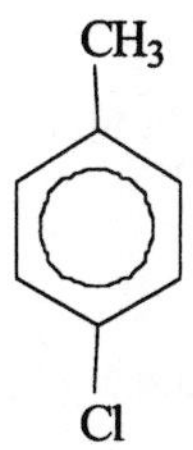 CH_3 Cl

Answers

1. benzene 2. bromobenzene 3. methylbenzene; toluene
4. 1,2-dichlorobenzene; *o*-dichlorobenzene
5. 1,3-dichlorobenzene; *m*-dichlorobenzene
6. nitrobenzene 7. 3,4-dichlorotoluene 8. 4-chlorotoluene; *p*-chlorotoluene

13.7 Properties of Aromatic Compounds

- Aromatic compounds have higher melting and boiling points than cycloalkanes.
- Aromatic compounds undergo substitution reactions of halogenation, nitration, and sulfonation.

◆ Learning Exercise 13.7

Write the missing reactant, catalyst, or product for each of the following reactions.

a. Benzene and $Br_2 \xrightarrow{FeBr_2}$

b. Benzene and $SO_3 \xrightarrow{H_2SO_4}$

c. Benzene and $HNO_3 \xrightarrow{H_2SO_4}$

d. 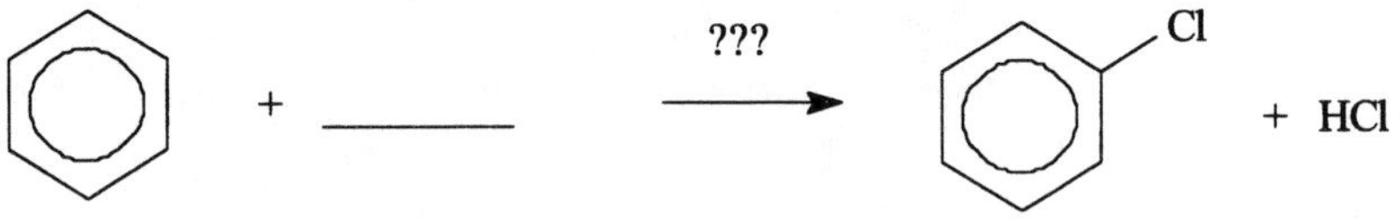 Benzene + ______ $\xrightarrow{???}$ chlorobenzene (Cl) + HCl

Answers

a. 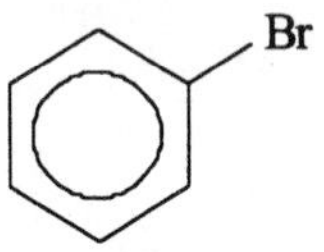 bromobenzene (Br) **b.** 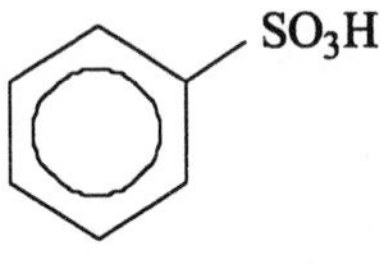 benzenesulfonic acid (SO_3H) **c.** 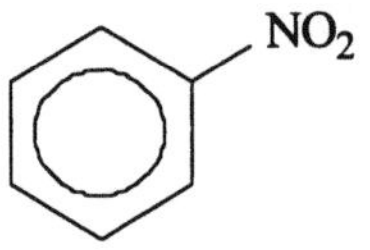 nitrobenzene (NO_2)

d. 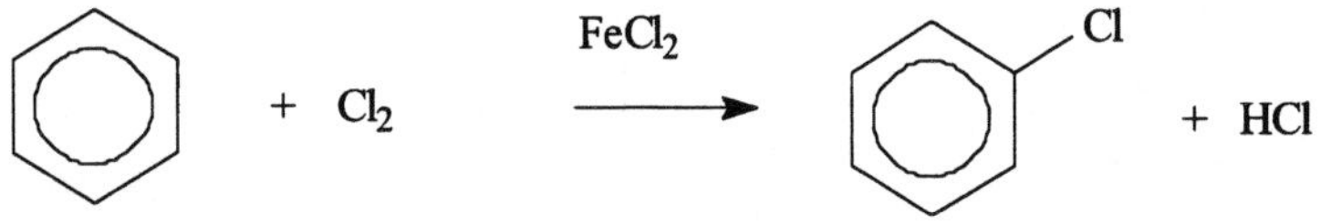 Benzene + Cl_2 $\xrightarrow{FeCl_2}$ chlorobenzene (Cl) + HCl

Check List for Chapter 13

You are ready to take the practice test for Chapter 13. Be sure that you have accomplished the following learning goals for this chapter. If you are not sure, review the section listed at the end of the goal. Then apply your new skills and understanding to the practice test.

After studying Chapter 13, I can successfully:

____ Identify the structural features of alkenes and alkynes (13.1).

____ Name alkenes and alkynes using IUPAC rules and write their structural formulas (13.2).

____ Identify alkenes that exist as cis–trans isomers; write their structural formulas and names (13.3).

____ Write the structural formulas and names for the products of the addition of hydrogen, halogens, hydrogen halides, and water to alkenes applying Markovnikov's rule when necessary (13.4).

____ Describe the process of forming polymers from alkene monomers (13.5).

____ Write the names and structures for compounds that contain a benzene ring (13.6).

____ Write the products of substitution reactions of benzene (13.7).

Practice Test for Chapter 13

Questions 1 through 4 refer to $H_2C{=}CH{-}CH_3$ (A) and $H_2C{-}CH_2{-}CH_2$ (cyclic, B).

1. These compounds are

   A. aromatic. **B.** alkanes. **C.** isomers.
   D. alkenes. **E.** cycloalkanes.

2. Compound (A) is a(n)

   A. alkane. **B.** alkene. **C.** cycloalkane.
   D. alkyne. **E.** aromatic.

3. Compound (B) is named

   A. propane. **B.** propylene. **C.** cyclobutane.
   D. cyclopropane. **E.** cyclopropene.

4. Compound (A) is named

   A. propane. **B.** propene. **C.** 2-propene.
   D. propyne. **E** 1-butene.

In questions 5 through 8, match the name of the alkene with the structural formula.

A. cyclopentene **B.** methylpropene **C.** cyclohexene
D. ethene **E.** 3-methylcyclopentene

5. $CH_2{=}CH_2$

6. $CH_3{-}C(CH_3){=}CH_2$

7. 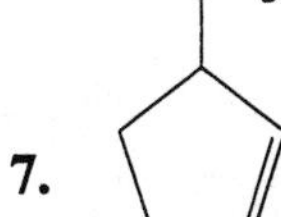

8. 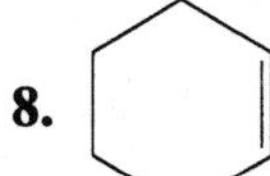

9. The *cis* isomer of 2-butene is

   A. $CH_2{=}CH{-}CH_2{-}CH_3$ **B.** $CH_3{-}CH{=}CH{-}CH_3$

   C. $(CH_3)(H)C{=}C(H)(CH_3)$ (CH$_3$ upper left, H upper right; H lower left, CH$_3$ lower right)

   D. $(CH_3)(H)C{=}C(CH_3)(H)$ (CH$_3$ upper left, CH$_3$ upper right; H lower left, H lower right)

   E. CH_3 CH_3 attached to $CH{=}CH$

10. The name of this compound is:

    $(Cl)(H)C{=}C(H)(Cl)$ (Cl upper left, H upper right; H lower left, Cl lower right)

    A. dichloroethene **B.** *cis*-1,2-dichloroethene **C.** *trans*-1,2-dichloroethene
    D. *cis*-chloroethene **E.** *trans*-chloroethene

11. Hydrogenation of $CH_3-CH=CH_2$ gives

A. $3CO_2 + 6H_2$ B. $CH_3-CH_2-CH_3$. C. $CH_2=CH-CH_3$.
D. no reaction. E. $CH_3-CH_2-CH_2-CH_3$.

12. The product of the reaction is: $CH_3-CH=CH_2 + HBr \longrightarrow$

A. $CH_3-CH_2-CH_2-Br$ B. no reaction C. $CH_3-CH_2-CH_3$

D. $CH_3-CH(Br)-CH_2-CH_2-Br$ E. $CH_3-CH(Br)-CH_3$

13. Addition of bromine (Br_2) to ethene gives

A. CH_3-CH_2-Br. B. $Br-CH_2-CH_2-Br$. C. $CH_3-CH-Br_2$.
D. CH_3-CH_3. E. no reaction.

14. Hydration of 2-butene gives

A. $CH_3-CH_2-CH_2-CH_3$ B. $CH_3-CH_2-CH_2-CH_2-OH$

C. $CH_3-CH(OH)-CH_2-CH_3$ D. □ E. □—OH

15. What is the common name for the compound 1,3-dichlorobenzene?

A. *m*-dichlorobenzene B. *o*-dichlorobenzene C. *p*-dichlorobenzene
D. *x*-dichlorobenzene E. *z*-dichlorobenzene

16. What is the common name of methylbenzene?

A. aniline B. phenol C. toluene
D. xylene E. toluidine

17. What is the IUPAC name of $CH_3-CH_2-C\equiv CH$?

A. methylacetylene B. propyne C. propylene
D. 4-butyne E. 1-butyne

18. What is the product when cyclopentene reacts with Cl_2?

A. chlorocyclopentene B. 1,1-dichlorocyclopentane C. 1,2-dichlorocyclopentane
D. 1,3-dichlorocyclopentane E. no reaction

19. The reaction $CH_2=CH_2 + Cl_2 \longrightarrow Cl-CH_2-CH_2-Cl$ is called

A. hydrogenation. B. halogenation. C. hydrohalogenation.
D. hydration. E. combustion.

20. The reaction in problem 19 is

A. a hydration reaction. B. an oxidation reaction. C. a substitution reaction.
D. an addition reaction. E. a reduction reaction.

21. The reaction $CH_3-CH=CH_2 + H_2O \longrightarrow CH_3-CH(OH)-CH_3$ is called a

A. hydrogenation. B. halogenation. C. hydrohalogenation of an alkene
D. hydration of an alkene. E. combustion.

For questions 22 through 25 identify the family for each compound as

A. alkane **B.** alkene **C.** alkyne **D.** cycloalkene

22. $CH_3—CH{=}CH_2$

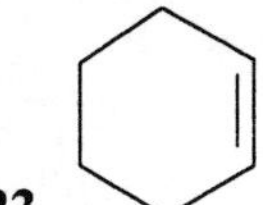

23.

24. $CH_3—CH_2—CH(CH_3)—CH_2—CH_3$

25. $CH_3—CH_2—C{\equiv}CH$

Match the name of each of the following aromatic compounds with the correct structure.

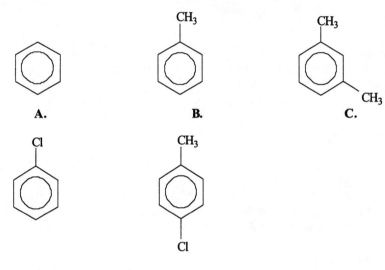

A.

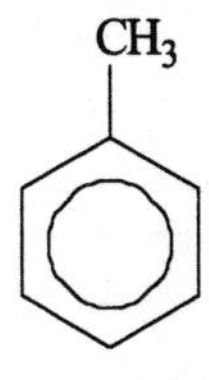

B.

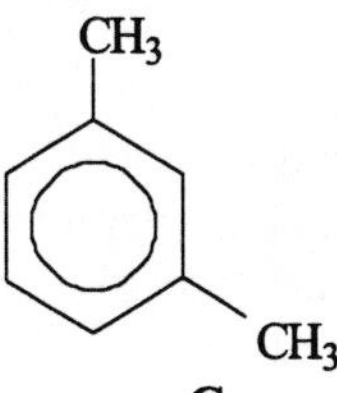

C.

Cl

D.

CH3

Cl

E.

26. ____ chlorobenzene

27. ____ benzene

28. ____ toluene

29. ____ *p*-chlorotoluene

30. ____ 1,3-dimethylbenzene

Answers to the Practice Test

1. C	2. B	3. D	4. B	5. D
6. B	7. E	8. C	9. D	10. C
11. B	12. E	13. B	14. C	15. A
16. C	17. E	18. C	19. B	20. D
21. D	22. B	23. D	24. A	25. C
26. D	27. A	28. B	29. E	30. C

Answers and Solutions to Selected Text Problems

13.1
- **a.** An alkane has only sigma bonds.
- **b.** An alkyne has a sigma bond and two pi bonds.
- **c.** An alkene has the groups on the carbon atom arranged at 120°.

13.3.
- **a.** An alkane has the general formula C_nH_{2n+2}.
- **b.** An alkene has a double bond.
- **c.** An alkyne has a triple bond.
- **d.** An alkene has a double bond.
- **e.** A cycloalkene has a double bond in a ring.

13.5. $CH_2{=}CH{-}CH_3$ Propene (propylene)

 Cyclopropane

13.7
- **a.** Propene contains three carbon atoms with a carbon-carbon double bond. Propyne contains three carbon atoms with a carbon-carbon triple bond.
- **b.** Cyclohexane is a six-carbon cyclic compound with all carbon–carbon single bonds. Cyclohexene is a six-carbon cyclic compound with a carbon–carbon double bond.

13.9
- **a.** The two-carbon compound with a double bond is ethene.
- **b.** 2-methyl-1-propene
- **c.** 4-bromo-2-pentyne
- **d.** There is a four-carbon cyclic structure with a double bond. The name is cyclobutene.
- **e.** There is a five-carbon cyclic structure with a double bond and an ethyl group. You must count the two carbons of the double bond as 1 and 2. The name is 4-ethylcyclopentene.
- **f.** Count the chain from the end nearest the double bond: 4-ethyl-2-hexene

13.11
- **a.** Propene is the three-carbon alkene. $H_2C{=}CH{-}CH_3$
- **b.** 1-pentene is the five-carbon compound with a double bond between carbonl and carbon 2.

  $H_2C{=}CH{-}CH_2{-}CH_2{-}CH_3$
- **c.** 2-methyl-1-butene has a four-carbon chain with a double bond between carbon 1 and carbon 2 and a methyl attached to carbon 2.

  $$\begin{array}{l} \quad\;\; CH_3 \\ \quad\;\;\; | \\ H_2C{=}C{-}CH_2{-}CH_3 \end{array}$$
- **d.** 3-methylcyclohexene is a six-carbon cyclic compound with a double bond between carbon 1 and carbon 2 and a methyl group attached to carbon 3.

  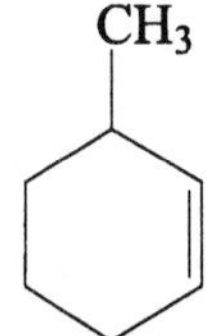

- **e.** 2-chloro-3-hexyne is a six-carbon compound with a triple bond between carbon 3 and 4 and a chlorine atom bonded to carbon 2.

  $$\begin{array}{l} \qquad\;\; Cl \\ \qquad\;\;\, | \\ CH_3{-}CH{-}C{\equiv}C{-}CH_2{-}CH_3 \end{array}$$

13.13 Constitutional isomers have the same number of each atom but the atoms are connected differently. Geometric isomers have different arrangements of the groups attached to a double bond.

13.15 There are four constitutional isomers with the molecular formula C_3H_5Cl; three are alkenes and one is a cycloalkane.

$Cl{-}CH_2{-}CH{=}CH_2$ $\qquad$ $CH_3{-}C(Cl){=}CH_2$

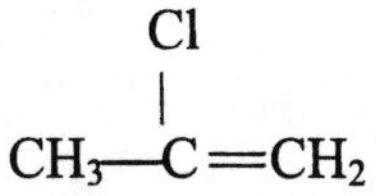

$CH_3{-}CH{=}CH{-}Cl$ $\qquad$ (cyclopropane ring bearing Cl)

13.17 **a.** This compound cannot have cis–trans isomers since there are two identical hydrogen atoms attached to the first carbon.
b. This compound can have cis–trans isomers since there are different groups attached to each carbon atom in the double bond.
c. This compound cannot have cis–trans isomers since there are two of the same groups attached to each carbon.

13.19 **a.** *cis*-2-butene. This is a four-carbon compound with a double bond between carbon 2 and carbon 3. Both methyl groups are on the same side of the double bond; it is cis.
b. *trans*-3-octene This compound has eight carbons with a double bond between carbon 3 and carbon 4. The alkyl groups are on opposite sides of the double bond; it is trans.
c. *cis*-3-heptene. This is a seven-carbon compound with a double bond between carbon 3 and carbon 4. Both alkyl groups are on the same side of the double bond; it is cis.

13.21 **a.** *trans*-2-butene has a four-carbon chain with a double bond between carbon 2 and carbon 3. The trans isomer has the two methyl groups on opposite sides of the double bond.

$$\begin{array}{ccc} CH_3 & & H \\ & C{=}C & \\ H & & CH_3 \end{array}$$

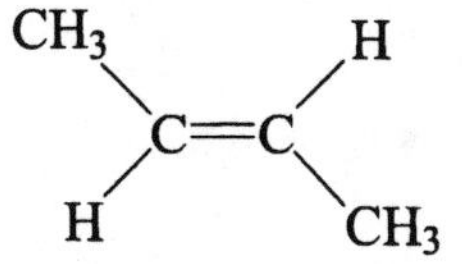

b. *cis*-2-pentene has a five-carbon chain with a double bond between carbon 2 and carbon 3. The cis isomer has the alkyl groups on the same side of the double bond.

$$\begin{array}{ccc} CH_3 & & CH_2{-}CH_3 \\ & C{=}C & \\ H & & H \end{array}$$

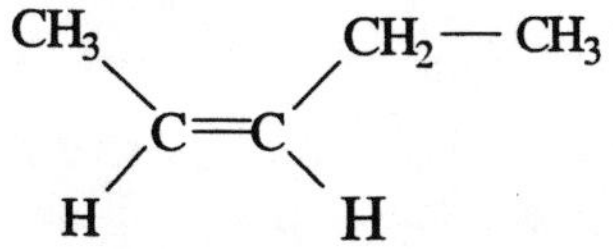

c. *trans*-3-heptene has a seven-carbon chain with a double bond between carbon 3 and carbon 4. The trans isomer has the alkyl groups on opposite sides of the double bond.

$$\begin{array}{ccc} CH_3{-}CH_2 & & H \\ & C{=}C & \\ H & & CH_2{-}CH_2{-}CH_3 \end{array}$$

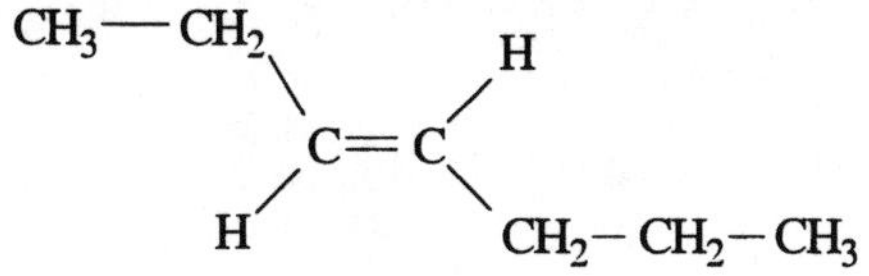

13.23 **a.** $CH_3{-}CH_2{-}CH_2{-}CH_2{-}CH_3$ $\qquad$ pentane

b. 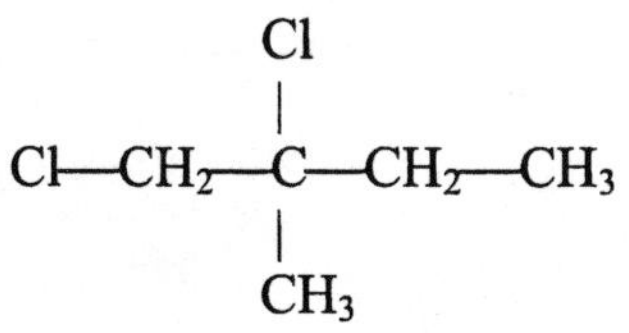 $Cl{-}CH_2{-}C(Cl)(CH_3){-}CH_2{-}CH_3$ $\qquad$ 1,2-dichloro-2-methylbutane

c. The product is a four-carbon cycloalkane with bromine atoms attached to carbon 1 and carbon 2. The name is 1,2-dibromocyclobutane.

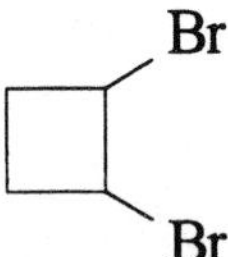

d. When H_2 is added to a cycloalkene the product is a cycloalkane. Cyclopentene would form cyclopentane.

+ H_2 $\xrightarrow{Pt}$

cyclopentene   cyclopentane

e. When Cl_2 is added to an alkene, the product is a dichloroalkane. The product is a four-carbon chain with chlorine atoms attached to carbon 2 and carbon 3 and a methyl group attached to carbon 2. The name of the product is: 2,3-dichloro-2-methylbutane.

$$CH_3-\overset{\displaystyle CH_3}{\overset{|}{C}}=CH-CH_3 + Cl_2 \longrightarrow CH_3-\underset{\displaystyle Cl}{\underset{|}{\overset{\displaystyle CH_3}{\overset{|}{C}}}}-\underset{\displaystyle Cl}{\underset{|}{CH}}-CH_3$$

2-methyl-2-butene   2,3-dichloro-2-methylbutane

f. $CH_3-CH_2-CH_2-CH_2-CH_3$ pentane

13.25 a. When HBr is added to an alkene, the product is a bromoalkane. In this case, we do not need to use Markovnikov's rule.

$$CH_3-CH_2-\overset{\displaystyle Br}{\overset{|}{CH}}-CH_3$$

b. When H_2O is added to an alkene, the product is an alcohol. In this case, we do not need to use Markovnikov's rule.

OH

c. When HCl is added to an alkene, the product is a chloroalkane. We need to use Markovnikov's rule, which says that the hydrogen that is added will go on the carbon that already has more hydrogens, in this case that is carbon 1.

$$CH_3-\overset{\displaystyle Cl}{\overset{|}{CH}}-CH_2-CH_3$$

d. When HI is added to an alkene, the product is a iodoalkane. In this case, we do not need to use Markovnikov's rule.

$$CH_3-\overset{\displaystyle CH_3}{\overset{|}{CH}}-\overset{\displaystyle I}{\overset{|}{C}}HCH_3$$

e. When HBr is added to an alkene, the product is a bromoalkane. We need to use Markovnikov's rule, which says that the hydrogen that is added will go on the carbon which already has more hydrogens, in this case that is carbon 2.

$$CH_3-CH_2-\underset{\underset{CH_3}{|}}{\overset{\overset{Br}{|}}{C}}-CH_2-CH_3$$

f. Using Markovnikov's rule, the H from HOH goes to the carbon 2 in the cyclohexane ring, which has more hydrogen atoms. The —OH then goes to carbon 1.

CH₃ / OH (1-methylcyclohexanol ring structure)

13.27 **a.** Hydrogenation of an alkene gives the saturated compound, the alkane.

$$CH_2=\overset{\overset{CH_3}{|}}{C}-CH_3 + H_2 \xrightarrow{Pt} CH_3-\overset{\overset{CH_3}{|}}{CH}-CH_3$$

b. The addition of HCl to a cycloalkene gives a chlorocycloalkane.

$$\text{cyclopentene} + HCl \longrightarrow \text{chlorocyclopentane (Cl)}$$

c. The addition of bromine (Br_2) to an alkene gives a dibromoalkane.

$$CH_3-CH=CH-CH_2-CH_3 + Br_2 \longrightarrow CH_3-\overset{\overset{Br}{|}}{CH}-\overset{\overset{Br}{|}}{CH}-CH_2-CH_3$$

d. Hydration (the addition of H_2O) to an alkene gives an alcohol. In this case, we use Markovnikov's rule and attach hydrogen to carbon 1.

$$CH_2=CH-CH_3 + H_2O \xrightarrow{H^+} CH_3-\overset{\overset{OH}{|}}{CH}-CH_3$$

e. $$CH_3-C\equiv C-CH_3 + 2Cl_2 \longrightarrow CH_3-\underset{\underset{Cl}{|}}{\overset{\overset{Cl}{|}}{C}}-\underset{\underset{Cl}{|}}{\overset{\overset{Cl}{|}}{C}}-CH_3$$

13.29 A polymer is a long-chain molecule consisting of many repeating smaller units. These smaller units are called monomers.

13.31 Teflon is a polymer of the monomer tetrafluoroethene.

$$3\ F-\overset{\overset{F}{|}}{C}=\overset{\overset{F}{|}}{C}-F \longrightarrow -\underset{\underset{F}{|}}{\overset{\overset{F}{|}}{C}}-\underset{\underset{F}{|}}{\overset{\overset{F}{|}}{C}}-\underset{\underset{F}{|}}{\overset{\overset{F}{|}}{C}}-\underset{\underset{F}{|}}{\overset{\overset{F}{|}}{C}}-\underset{\underset{F}{|}}{\overset{\overset{F}{|}}{C}}-\underset{\underset{F}{|}}{\overset{\overset{F}{|}}{C}}-$$

13.33 1,1-difluoroethene is the two carbon alkene with two fluorine atoms attached to carbon 1.

```
    F   H   F   H   F   H
    |   |   |   |   |   |
  —C— C— C— C— C— C—
    |   |   |   |   |   |
    F   H   F   H   F   H
```

13.35 Cyclohexane, C_6H_{12}, is a cycloalkane in which six carbon atoms are linked by single bonds in a ring. In benzene, C_6H_6, an aromatic system links the six carbon atoms in a ring.

13.37 The six-carbon ring with alternating single and double bonds is benzene. If the groups are in the 1,2 position, this is ortho (*o*), 1,3 is meta (*m*), and 1,4 is para (*p*).

a. 1-chloro-2-methylbenzene; *o*-chlorotoluene
b. ethylbenzene
c. 1,3,5-trichlorobenzene
d. *m*-xylene; *m*-methyltoluene; 1,3-dimethylbenzene
e. 1-bromo-3-chloro-5-methylbenzene; 3-bromo-5-chlorotoluene
f. isopropyl benzene

13.39 **a.** CH_3

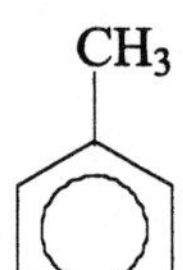

b. Cl Cl

The prefix *m* means that the two chloro groups are in the 1 and 3 position.

c. CH_2CH_3

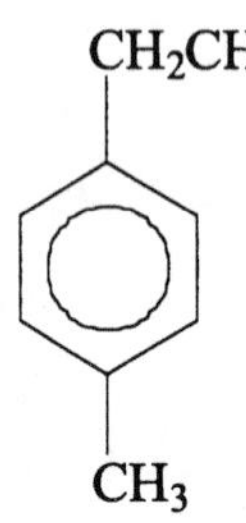

CH_3

d. CH_3

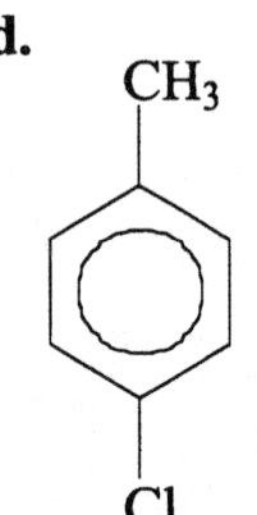

Cl

The prefix *p* means that the two groups are in the 1 and 4 position.

13.41 Benzene undergoes substitution reactions because a substitution reaction allows benzene to retain the stability of the aromatic system.

13.43 **a.** No reaction

b. Cl

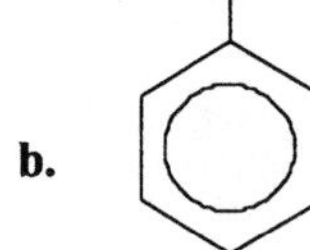

c. NO_2

13.45 Propane is the three-carbon alkane with the formula C_3H_8. All the carbon–carbon bonds in propane are single bonds. Cyclopropane is the three carbon cycloalkane with the formula C_3H_6. All the carbon–carbon bonds in cyclopropane are single bonds. Propene is the three-carbon compound which has a carbon–carbon double bond. The formula of propene is C_3H_6. Propyne is the three-carbon compound with a carbon–carbon triple bond. The formula of propyne is C_3H_4.

13.47 **a.** This compound has a chlorine atom attached to a cyclopentane; the IUPAC name is chlorocyclopentane.
b. This compound has a five-carbon chain with a chlorine atom attached to carbon 2 and a methyl group attached to carbon 4. The IUPAC name is: 2-chloro-4-methylpentane.

c. This compound contains a five-carbon chain with a double bond between carbon 1 and carbon 2 and a methyl group attached to carbon 2. The IUPAC name is: 2-methyl-1-pentene.

d. This compound contains a five-carbon chain with a triple bond between carbon 2 and carbon 3. The IUPAC name is: 2-pentyne.

e. This compound contains a five-carbon cycloalkene with a chlorine atom attached to carbon 1. The IUPAC name is: 1-chlorocyclopentene.

f. This compound contains a five-carbon chain with a double bond between carbon 2 and carbon 3. The alkyl groups are on opposite sides of the double bond. The IUPAC name is: *trans*-2-pentene.

g. This compound contains a six-carbon ring with a double bond and chlorine atoms attached to carbon 1 and carbon 3. The IUPAC name is: 1,3-dichlorocyclohexene.

13.49 **a.** These structures represent a pair of constitutional isomers. In one isomer, the chlorine is attached to one of the carbons in the double bond; in the other isomer, the carbon bonded to the chlorine is not part of the double bond.

b. These structures are cis–trans isomers. In the cis isomer, the two methyl groups are on the same side of the double bond. In the trans isomer, the methyl groups are on opposite sides of the double bond.

c. These structures are identical and not isomers. Both have five-carbon chains with a double bond between carbon 1 and carbon 2.

d. These structures represent a pair of constitutional isomers. Both have the molecular formula C_7H_{16}. One isomer has a six-carbon chain with a methyl group attached, the other has a five-carbon chain with two methyl groups attached.

13.51 The structure of methylcyclopentane is

CH_3

It can be formed by the hydrogenation of four cycloalkenes.

CH_3 CH_3 CH_3 CH_2

13.53 **a.**

CH_3 CH_2CH_3
C=C
H H

cis-2-pentene; both alkyl groups are on the same side of the double bond

CH_3 H
C=C
H CH_2CH_3

trans-2-pentene; both alkyl groups are on opposite sides of the double bond

b.

CH_3CH_2 CH_2CH_3
C=C
H H

cis-3-hexene; both alkyl groups are on the same side of the double bond

CH_3CH_2 H
C=C
H CH_2CH_3

trans-3-hexene; both alkyl groups are on opposite sides of the double bond

c. CH_3 / CH_3 — $C{=}C$ — H / H

cis-2-butene; both alkyl groups are on the same side of the double bond

CH_3 / H — $C{=}C$ — H / CH_3

trans-2-butene; both alkyl groups are on opposite sides of the double bond

d. CH_3 / $CH_2CH_2CH_3$ — $C{=}C$ — H / H

cis-2-hexene; both alkyl groups are on the same side of the double bond

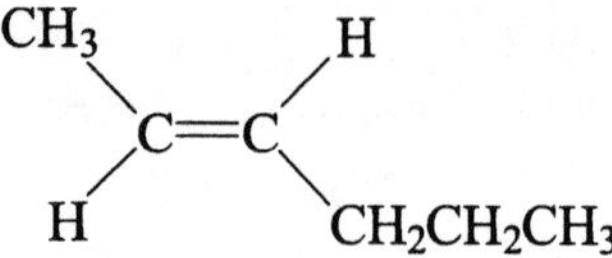

trans-2-hexene; both alkyl groups are on opposite sides of the double bond

13.55 **a.** The reaction of H_2 in the presence of a Ni catalyst changes alkenes into alkanes. The reactant must be cyclohexene.

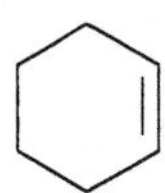

b. Br_2 adds to alkenes to give a dibromoalkane. Since there are bromine atoms on carbon 2 and carbon 3 the double bond in the reactant must have been between carbons 2 and 3.

$CH_3CH{=}CHCH_2CH_3$

c. HCl adds to alkenes to give a chloroalkane. The product has three carbons and the double bond must be between carbons 1 and 2.
$CH_2{=}CHCH_3$

d. An alcohol is formed when H_2O adds to an alkene in the presence of acid (H^+). The alkene which adds water to form

$$\text{cyclopentene} + H_2O \xrightarrow{H^+} \text{cyclopentanol (OH)}$$

13.57 Styrene is $H_2C{=}CH$–(benzene ring) and acrylonitrile is $H_2C{=}CH$–CN. A section copolymer of styrene and acrylonitrile would be:

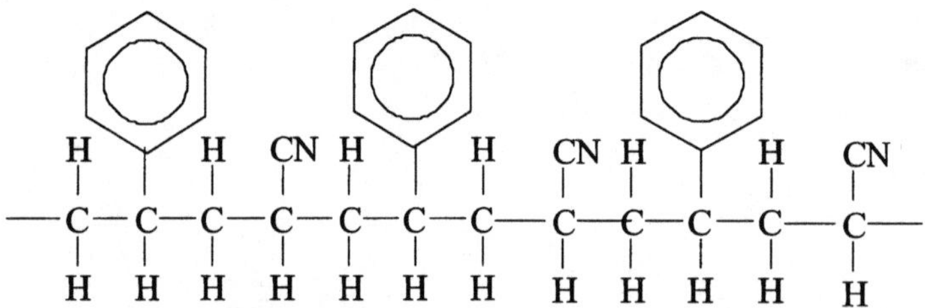

13.59 **a.** methylbenzene; toluene
b. 1-chloro-2-methylbenzene; *o*-chlorotoluene (1,2 position is *ortho, o*)
c. 1-ethyl-4-methylbenzene; *p*-ethyltoluene (1,4 position is *para, p*)
d. 1,3-diethylbenzene; *m*-diethylbenzene (1,3 position is *meta, m*)

13.61 **a.** Chlorobenzene
b. *o*-bromotoluene, *m*-bromotoluene, *p*-bromotoluene
c. Benzenesulfonic acid
d. No products

14

Alcohols, Phenols, Ethers, and Thiols

Study Goals

- Classify alcohols as primary, secondary, or tertiary.
- Name and write the condensed structural formulas for alcohols, phenols, and thiols.
- Identify the uses of some alcohols and phenols.
- Name and write the condensed structural formulas for ethers.
- Describe the solubility in water, density, and boiling points of alcohols, phenols, and ethers.
- Write equations for combustion, dehydration, and oxidation of alcohols.

Think About It

1. What are the functional groups of alcohols, phenols, ethers, and thiols?

2. Phenol is sometime used in mouthwashes. Why does it form a solution with water?

3. What reaction of ethanol takes place when you make a fondue dish or a flambé dessert?

Key Terms

Match the following terms with the statements shown below.

a. primary alcohol **b.** thiol **c.** ether **d.** phenol **e.** tertiary alcohol

1. _____ An organic compound with one alkyl group bonded to the carbon with the —OH group

2. _____ An organic compound that contains an —SH group

3. _____ An organic compound that contains an oxygen atom —O— attached to two alkyl groups

4. _____ An organic compound with three alkyl groups bonded to the carbon with the —OH group

5. _____ An organic compound that contains a benzene ring bonded to a hydroxyl group

Answers **1.** a **2.** b **3.** c **4.** e **5.** d

14.1 Structure and Classification of Alcohols

- Alcohols are classified according to the number of alkyl groups attached to the carbon bonded to the —OH group.
- In a primary alcohol, there is one alkyl group attached to the carbon atom bonded to the —OH. In a secondary alcohol, there are two alkyl groups, and in a tertiary alcohol there are three alkyl groups attached to the carbon atom with the —OH functional group.

Study Note

Example: Identify the following as primary, secondary, or tertiary alcohols.
Solution: Determine the number of alkyl groups attached to the hydroxyl carbon atom.

$CH_3—CH_2—OH$	$CH_3—CH(CH_3)—OH$	$CH_3—C(CH_3)_2—OH$
Primary (1°)	Secondary (2°)	Tertiary (3°)

◆ **Learning Exercise 14.1**

Classify each of the following alcohols as primary (1°), secondary (2°), or tertiary (3°).

1. $CH_3—CH_2—OH$ ____
2. $CH_3—CH_2—CH(OH)—CH_3$ ____
3. $CH_3—C(OH)(CH_3)—CH_2—CH_3$ ____
4. $CH_3—C(OH)(CH_3)—CH_2—CH_2—CH_3$ ____
5. $CH_3—C(CH_3)_2—CH_2—OH$ ____
6. 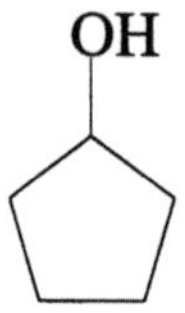 cyclopentane ring with OH ____

Answers **1.** primary (1°) **2.** secondary (3°) **3.** tertiary (3°) **4.** tertiary (3°) **5.** primary (1°) **6.** secondary (2°)

14.2 Naming Alcohols, Phenols, and Thiols

- An alcohol contains the hydroxyl group —OH attached to a carbon chain.
- In the IUPAC system alcohols are named by replacing the *ane* of the alkane name with *ol*. The location of the —OH group is given by numbering the carbon chain. Simple alcohols are generally named by their common names with the alkyl name preceding the term alcohol. For example, $CH_3—OH$ is methyl alcohol, and $CH_3—CH_2—OH$ is ethyl alcohol.

$CH_3—OH$	$CH_3—CH_2—OH$	$CH_3—CH_2—CH_2—OH$
methanol	ethanol	1-propanol
(methyl alcohol)	(ethyl alcohol)	(propyl alcohol)

- When a hydroxyl group is attached to a benzene ring, the compound is a phenol.
- Thiols are similar to alcohols, except they have an —SH functional group in place of the —OH group, R—SH. To name a thiol, give the alkane name of the chain followed by *thiol*.

$CH_3—SH$	$CH_3—CH_2—SH$
methanethiol	ethanethiol

◆ Learning Exercise 14.2A

Give the correct IUPAC and common name (if any) for each of the following compounds.

1. $CH_3—CH_2—OH$

2. $CH_3—CH_2—CH_2—OH$

3. 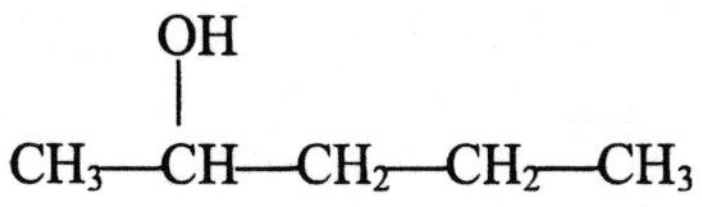 $CH_3—CH(OH)—CH_2—CH_2—CH_3$

4. 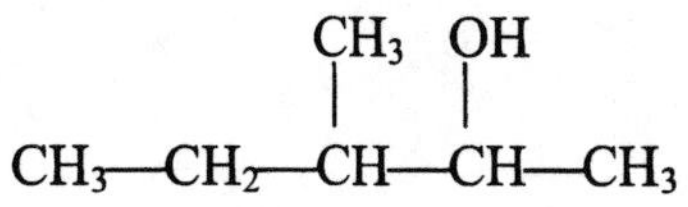 $CH_3—CH_2—CH(CH_3)—CH(OH)—CH_3$

5. 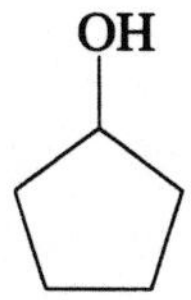 (cyclopentane ring with OH)

6. 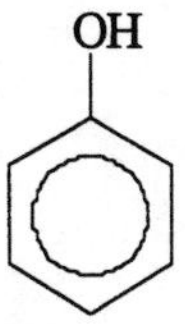 (benzene ring with OH)

Answers

1. ethanol (ethyl alcohol)
2. 1-propanol (propyl alcohol)
3. 2-pentanol
4. 3-methyl-2-pentanol
5. cyclopentanol
6. phenol

◆ Learning Exercise 14.2B

Write the correct condensed structural formula for each of the following compounds.

1. 2-butanol

2. 2-chloro-1-propanol

3. 2,4-dimethyl-1-pentanol

4. cyclohexanol

5. 3-methylcyclopentanol

6. *o*-chlorophenol

Answers

1. $CH_3—\underset{|}{\overset{OH}{CH}}—CH_2—CH_3$

2. $CH_3—\underset{|}{\overset{Cl}{CH}}—CH_2—OH$

3. $CH_3—\overset{CH_3}{CH}—CH_2—\overset{CH_3}{CH}—CH_2—OH$

4. 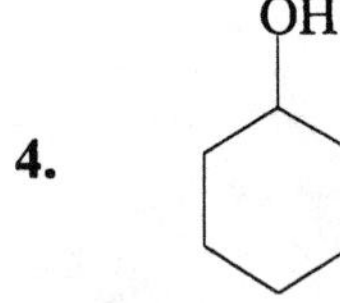

5. OH, CH₃ substituted cyclopentane (3-methylcyclopentanol)

6. OH, Cl substituted benzene ring

◆ Learning Exercise 14.2C

Give the correct IUPAC name for the following thiols.

1. $CH_3—CH_2—SH$ ____________________________
2. $CH_3—CH_2—CH_2—SH$ ____________________________
3. $CH_3—CH_2—\overset{SH}{CH}—CH_3$ ____________________________
4. SH on cyclobutane ring ____________________________

Answers **1.** ethanethiol **2.** 1-propanethiol **3.** 2-butanethiol **4.** cyclobutanethiol

14.3 Some Important Alcohols and Phenols

- Methanol $CH_3—OH$, the simplest alcohol, and ethanol, $CH_3—CH_2—OH$ are found in many solvents and paint removers.
- Glycerol, a trihydroxy alcohol, is obtained from oils and fats and used as a skin softener.
- Derivatives of phenol such as resorcinol are used as antiseptics in throat lozenges and mouthwashes.

◆ Learning Exercise 14.3

Identify the alcohol or phenol that matches each of the following descriptions.

1. Used to make alcoholic beverages ______________________
2. Used to clean the skin before giving an injection or drawing a blood sample ______________________
3. Forms an insoluble salt that damages the kidneys ______________________

Answers **1.** ethanol **2.** isopropyl alcohol **3.** 1,2-ethanediol (ethylene glycol)

14.4 Ethers

- In ethers, an oxygen atom is connected by single bonds to two alkyl or aromatic groups, R—O—R.
- In the IUPAC name, the smaller alkyl group and the oxygen are named as an *alkoxy group* attached to the longer alkane chain, which is numbered to give the location of the alkoxy group. In the common names of ethers, the alkyl groups are listed alphabetically followed by the name *ether*.

Study Note

Example: Write the common and IUPAC names for CH_3—CH_2—O—CH_3.

Solution: The common name lists the alkyl groups alphabetically before the name *ether* Using the IUPAC system, the smaller alkyl group and the oxygen are named as a substituent *methoxy* attached to the two-carbon chain ethane.

ethyl group *methyl group*	*ethane* *methoxy group*
CH_3—CH_2—O—CH_3	CH_3—CH_2—O—CH_3
Common: ethyl methyl ether	IUPAC: methoxy ethane

◆ Learning Exercise 14.4A

Write an IUPAC and common name, if any, for the following ethers.

1. CH_3—O—CH_3

2. CH_3—CH_2—O—CH_2—CH_3

3. CH_3—CH_2—CH_2—CH_2—O—CH_3

4. CH_3—O—CH_2—CH_3

5. (benzene ring)—OCH_3

Answers

1. methoxymethane; (di)methyl ether

2. ethoxyethane; (di)ethyl ether

3. 1-methoxybutane; butyl methyl ether

4. methoxyethane; ethyl methyl ether

5. methoxybenzene; methyl phenyl ether (anisole)

◆ Learning Exercise 14.4B

Write the structural formula for each of the following ethers.

1. ethyl propyl ether

2. 2-methoxypropane

3. ethyl methyl ether

4. 3-ethoxypentane

Answers

1. CH_3—CH_2—O—CH_2—CH_2—CH_3

2.
```
         O—CH3
         |
   CH3—CH—CH3
```

3. CH_3—O—CH_2—CH_3

4.
```
               O—CH2—CH3
               |
   CH3—CH2—CH—CH2—CH3
```

◆ Learning Exercise 14.4C

Identify each of the following heterocyclic structures as a furan, pyran, or dioxane.

1. 

2.

3. 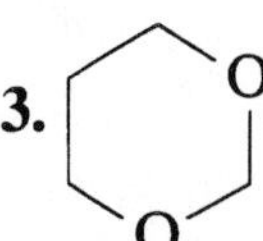

4. 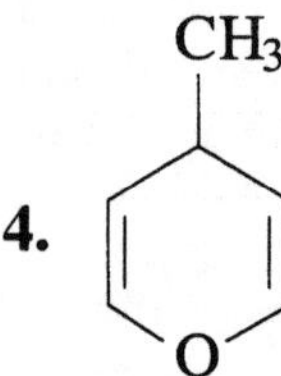

Answers **1.** furan **2.** pyran **3.** dioxane **4.** pyran

◆ Learning Exercise 14.4D

Identify each of the following pairs of compounds as constitutional isomers, conformational isomers of the same compound, or different compounds.

1. CH_3—O—CH_3 and CH_3—CH_2—OH ____________

```
                                    OH
                                    |
2. CH3—O—CH2—CH3 and CH3—CH—CH3
```
____

```
        CH3                 OH
         |                   |
3. CH3—CH—OH and CH3—CH—CH3
```
____

4. CH_3—CH_2—O—CH_3 and CH_3—CH_2—CH_2—CH_2—OH ____________

Answers **1.** constitutional isomers **2.** constitutional isomers
3. conformation isomers **4.** different compounds

14.5 Physical Properties of Alcohols, Phenols, and Ethers

- The polar —OH group gives alcohols higher boiling points than alkanes and ethers of similar mass.
- Alcohols with one to four carbons are soluble in water because the —OH group forms hydrogen bonds with water molecules.
- Phenol is soluble in water and acts as a weak acid.
- Because ethers are less polar than alcohols, they have boiling points similar to alkanes. Ethers are soluble in water due to hydrogen bonding. Ethers are widely used as solvents but can be dangerous to use because their vapors are highly flammable.

◆ Learning Exercise 14.5A

Circle the compound in each pair that is the more soluble in water.

1. ____ CH_3—CH_3 or CH_3—CH_2—OH

2. ____ CH_3—CH_2—CH_2—OH or CH_3—CH_2—CH_2—CH_2—CH_2—OH

3. ____ CH_3—CH_2—CH_2—CH_3 or CH_3—CH_2—CH_2—CH_2—OH

4. ____ Benzene or phenol

Answers

1. $CH_3—CH_2—OH$
2. $CH_3—CH_2—CH_2—OH$
3. $CH_3—CH_2—CH_2—CH_2—OH$
4. phenol

◆ Learning Exercise 14.5B

Select the compound in each pair with the higher boiling point.

1. $CH_3—CH_2—CH_3$ or $CH_3—CH_2—OH$
2. 2-butanol or 2-hexanol
3. $CH_3—O—CH_2—CH_3$ or $CH_3—CH_2—CH_2—OH$
4. $CH_3—CH_2—CH_2—OH$ or $CH_3—CH_2—CH_2—CH_3$

Answers

1. $CH_3—CH_2—OH$
2. 2-hexanol
3. $CH_3—CH_2—CH_2—OH$
4. $CH_3—CH_2—CH_2—OH$

14.6 Reactions of Alcohols

- At high temperatures, an alcohol dehydrates in the presence of an acid to yield an alkene and water.

$$CH_3—CH_2—OH \xrightarrow[\text{Heat}]{H^+} H_2C{=}CH_2 + H_2O$$

- Ethers are produced from primary alcohols in the presence of acid and at lower temperatures than needed for dehydration.

$$CH_3—OH + HO—CH_3 \xrightarrow[\text{Heat}]{H^+} CH_3—O—CH_3 + H_2O$$

- Using an oxidizing agent[O], primary alcohols oxidize to aldehydes, which usually oxidize further to carboxylic acids. Secondary alcohols are oxidized to ketones, but tertiary alcohols do not oxidize.

$$\underset{1°\text{ alcohol}}{CH_3—CH_2—OH} \xrightarrow{[O]} \underset{\text{aldehyde}}{CH_3—\overset{\overset{\displaystyle O}{\|}}{C}—H} + H_2O$$

$$\underset{2°\text{ alcohol}}{CH_3—\overset{\overset{\displaystyle OH}{|}}{CH}—CH_3} \xrightarrow{[O]} \underset{\text{ketone}}{CH_3—\overset{\overset{\displaystyle O}{\|}}{C}—CH_3} + H_2O$$

◆ Learning Exercise 14.6A

Write the condensed structural formulas of the products expected from dehydration of each of the following reactants:

1. $CH_3—CH_2—CH_2—CH_2—OH \xrightarrow{H^+,\ \text{heat}}$
2. (cyclohexanol: cyclohexane ring with OH) $\xrightarrow{H^+,\ \text{heat}}$

3. $$CH_3\text{—}\underset{\displaystyle OH}{\overset{|}{CH}}\text{—}CH_3 \xrightarrow{H^+,\ heat}$$

4. $$CH_3\text{—}CH_2\text{—}\underset{}{\overset{OH}{\overset{|}{CH}}}\text{—}CH_2\text{—}CH_3 \xrightarrow{H^+,\ heat}$$

Answers

1. $CH_3\text{—}CH_2\text{—}CH{=}CH_2$
2. 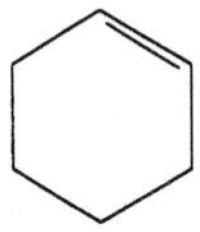 (cyclohexene structure)
3. $CH_3\text{—}CH{=}CH_2$
4. $CH_3\text{—}CH_2\text{—}CH{=}CH\text{—}CH_3$

◆ Learning Exercise 14.6B

Write the structure of the ether formed in the following reactions.

1. $CH_3\text{—}CH_2\text{—}OH + HO\text{—}CH_2\text{—}CH_3 \xrightarrow{H^+,\ heat}$
2. $CH_3\text{—}OH + HO\text{—}CH_3 \xrightarrow{H^+,\ heat}$

Answers 1. $CH_3\text{—}CH_2\text{—}O\text{—}CH_2\text{—}CH_3$ 2. $CH_3\text{—}O\text{—}CH_3$

◆ Learning Exercise 14.6C

Write the condensed structural formulas of the products expected in the oxidation reaction of each of the following reactants.

1. $CH_3\text{—}CH_2\text{—}CH_2\text{—}CH_2\text{—}OH \xrightarrow{[O]}$
2. 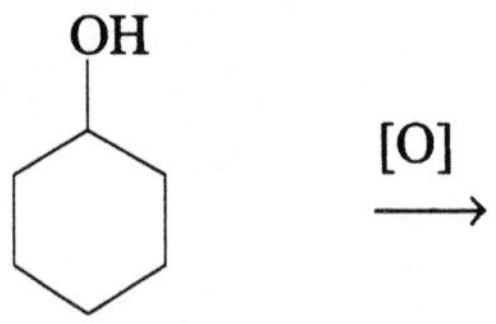 (cyclohexanol: cyclohexane ring with OH) $\xrightarrow{[O]}$
3. $CH_3\text{—}\overset{OH}{\overset{|}{CH}}\text{—}CH_3 \xrightarrow{[O]}$
4. $CH_3\text{—}CH_2\text{—}\overset{OH}{\overset{|}{CH}}\text{—}CH_2\text{—}CH_3 \xrightarrow{[O]}$

Answers

1. $CH_3—CH_2—CH_2—\overset{\overset{\displaystyle O}{\|}}{C}—H$

2. (cyclohexanone structure: six-membered ring with C=O)

3. $CH_3—\overset{\overset{\displaystyle O}{\|}}{C}—CH3$

4. $CH_3—CH_2—\overset{\overset{\displaystyle O}{\|}}{C}—CH_2—CH_3$

Check List for Chapter 14

You are ready to take the practice test for Chapter 14. Be sure that you have accomplished the following learning goals for this chapter. If you are not sure, review the section listed at the end of the goal. Then apply your new skills and understanding to the practice test.

After studying Chapter 14, I can successfully:

_____ Classify an alcohol as primary, secondary, or tertiary (14.1).

_____ Give the IUPAC or common name of an alcohol, phenol, or thiol; draw the condensed structural formula from the name. (14.2).

_____ Identify the uses of some important alcohols and phenols (14.3).

_____ Write the IUPAC or common name of an ether; write the condensed structural formula from the name (14.4).

_____ Describe the solubility of alcohols, phenols, and ethers in water; compare their boiling points (14.5).

_____ Write the products of alcohols that undergo dehydration, ether formation, and oxidation (14.6).

Practice Test for Chapter 14

Match the names of the following compounds with their structures.

A. 1-propanol **B.** cyclobutanol **C.** 2-propanol **D.** ethyl methyl ether **E.** diethyl ether

1. 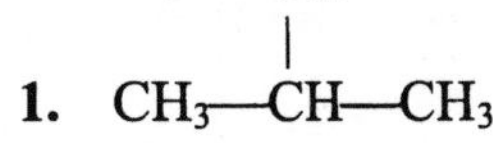 $CH_3—\overset{\overset{\displaystyle OH}{|}}{CH}—CH_3$

2. $CH_3—CH_2—CH_2—OH$

3. $CH_3—O—CH_2—CH_3$

4. (cyclobutanol structure: four-membered ring with OH)

5. $CH_3—CH_2—O—CH_2—CH_3$

6. The compound $CH_3—\overset{\overset{\displaystyle O}{\|}}{C}—CH_3$ is formed by the oxidation of

A. 2-propanol. **B.** propane. **C.** 1-propanol.
D. dimethyl ether. **E.** methyl ethyl ketone.

7. Why are short-chain alcohols water soluble?

A. They are nonpolar. **B.** They can hydrogen bond. **C.** They are organic.
D. They are bases. **E.** They are acids.

8. Phenol is

A. the alcohol of benzene. **B.** the aldehyde of benzene.
C. the phenyl group of benzene. **D.** the ketone of benzene. **E.** cyclohexanol.

9. $CH_3—CH_2—OH + HO—CH_2—CH_3 \xrightarrow{H^+}$ [ ] $+ H_2O$

A. an alkane **B.** an aldehyde **C.** a ketone
D. an ether **E.** a phenol

10. The dehydration of cyclohexanol gives

A. cyclohexane. **B.** cyclohexene. **C.** cyclohexyne.
D. benzene. **E.** phenol.

11. The formula of ethanethiol is

A. $CH_3—SH$. **B.** $CH_3—CH_2—OH$. **C.** $CH_3—CH_2—SH$.
D. $CH_3—CH_2—S—CH_3$ **E.** $CH_3—S—OH$.

In questions 12 through 16, classify each alcohol as

A. primary (1°) **B.** secondary (2°) **C.** tertiary (3°)

12. $CH_3—CH_2—CH_2—OH$

13. 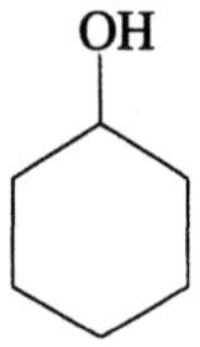 (cyclohexane ring with OH)

14. 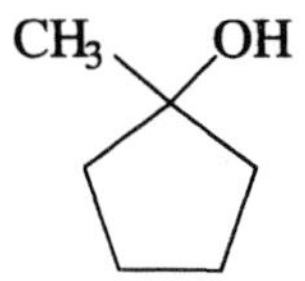 (cyclopentane ring with CH_3 and OH on the same carbon)

15. $CH_3—C(OH)(CH_3)—CH_2—CH_2—CH_3$

16. $CH_3—CH(OH)—CH_2—CH_2—CH_2—CH_3$

Complete questions 17 through 20 by indicating one of the products (A–E) formed in each of the following reactions.

A. primary alcohol **B.** secondary alcohol **C.** aldehyde **D.** ketone **E.** carboxylic acid

17. ____ Oxidation of a primary alcohol

18. ____ Oxidation of a secondary alcohol

19. ____ Oxidation of an aldehyde

20. ____ Hydration of 1-propene

Answers to Practice Test

1. C	**2.** A	**3.** D	**4.** B	**5.** E
6. A	**7.** B	**8.** A	**9.** D	**10.** B
11. C	**12.** A	**13.** B	**14.** C	**15.** C
16. B	**17.** C	**18.** D	**19.** E	**20.** B

Answers and Solutions to Selected Text Problems

14.1 In a primary (1°) alcohol, the carbon bonded to the hydroxyl group (—OH) is attached to one alkyl group (except for methanol); to two alkyl groups in a secondary alcohol (2°); and to three alkyl groups in a tertiary alcohol (3°).

a. 1° **b.** 1° **c.** 3° **d.** 2° **e.** 3°

14.3 a. This compound has a two-carbon chain (ethane). The final –e is dropped and –ol added to indicate an alcohol. The IUPAC name is ethanol.

b. This compound has a four-carbon chain with a hydroxyl attached to carbon 2. The IUPAC name is 2-butanol.

c. This compound has a six-carbon chain with a hydroxyl group attached to carbon 3. The IUPAC name is 3-hexanol.

d. This compound has a four-carbon chain with a hydroxyl attached to carbon 1and a methyl attached to carbon 3. The IUPAC name is 3-methyl-1-butanol.

e. This compound is a six-carbon cycloalkane with a hydroxyl attached to carbon 1 and two methyl groups, one attached to carbon 3 and the other to carbon 4. Since the hydroxyl is always attached to carbon 1 the number 1 is omitted in the name. The IUPAC name is 3,4-dimethylcyclohexanol.

f. This compound has a seven-carbon chain with a hydroxyl attached to carbon 1 and three methyl groups, one attached to carbon 3 and two attached to carbon 5. The IUPAC name is 3,5,5-trimethyl-1-heptanol.

14.5 a. 1-propanol has a three-carbon chain with a hydroxyl attached to carbon 1.

$CH_3—CH_2—CH_2—OH$

b. Methyl alcohol has a hydroxyl attached to a one-carbon alkane. $CH_3—OH$.

c. 3-pentanol has a five-carbon chain with a hydroxyl attached to carbon 3.

$$\begin{array}{c} \quad\quad OH \\ \quad\quad | \\ CH_3—CH_2—CH—CH_2—CH_3 \end{array}$$

d. 2-methyl-2-butanol has a four-carbon chain with a methyl and hydroxyl attached to carbon 2.

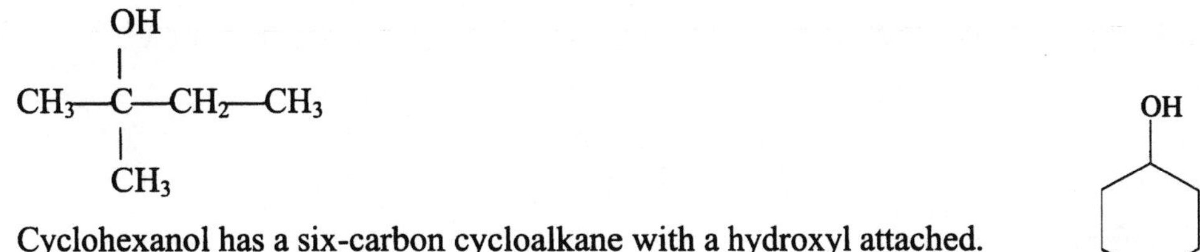

e. Cyclohexanol has a six-carbon cycloalkane with a hydroxyl attached.

f. 1,4-butanediol has a four-carbon chain (butane). The diol indicates that there are two hydroxyl groups, one attached to carbon 1 and one to carbon 4.

$HO—CH_2—CH_2—CH_2—CH_2—OH$

14.7 A benzene ring with a hydroxyl group is called *phenol.* Substituents are numbered from the carbon bonded to the hydroxyl group as carbon 1. Common names use *ortho, meta,* and *para* prefixes.

- **a.** phenol
- **b.** 2-bromophenol, *ortho*-bromophenol (Groups on the 1 and 2 positions are ortho)
- **c.** 3,5-dichlorophenol
- **d.** 3-bromophenol, *meta*-bromophenol (Groups on the 1 and 3 positions are meta)

14.9 **a.** The *m* (meta) indicates that the two groups are in the 1,3 arrangement.

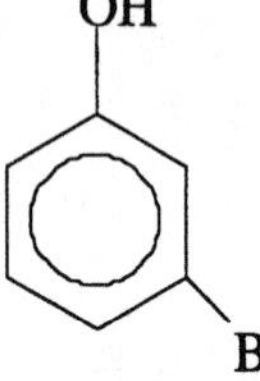

b. The *p* (para) indicates that the two groups are in the 1,4 arrangement.

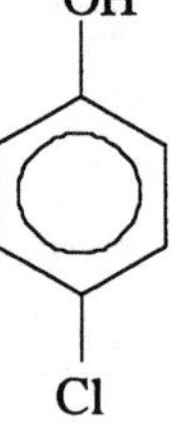

c. Two chlorine atoms are attached to the aromatic system, one on carbon 2 and the other on carbon 5, with the hydroxyl attached to carbon 1.

d. The *o* (ortho) indicates that the two groups are in the 1,2 arrangement.

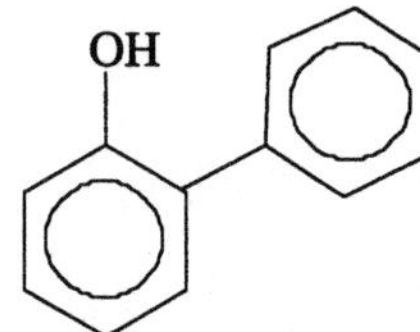

e. The name indicated that there is an ethyl group on carbon 4 with the hydroxyl on carbon 1.

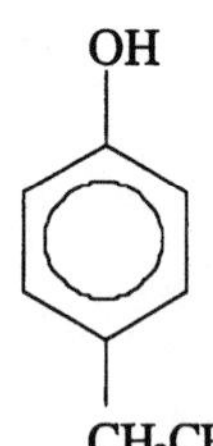

14.11 **a.** This is a one-carbon alkane with a thiol (—SH) group. The IUPAC name is methanethiol.
b. This thiol has a three-carbon alkane with the thiol group attached to carbon 2. The IUPAC name is 2-propanethiol.
c. This compound has a four-carbon alkane with methyl groups attached to carbon 2 and carbon 3 and the thiol attached to carbon 1. The IUPAC name is 2,3-dimethyl-1-butanethiol.
d. This compound has a thiol attached to a cyclobutane. The IUPAC name is cyclobutanethiol.

14.13 **a.** ethanol **b.** thymol **c.** *ortho*-phenylphenol

14.15 **a.** methoxyethane; ethyl methyl ether
b. methoxycyclohexane, cyclohexyl methyl ether
c. ethoxycyclobutane, cyclobutyl ethyl ether
d. 1-methoxypropane, methyl propyl ether

14.17 **a.** Ethyl propyl ether has a two-carbon group and a three-carbon group attached to oxygen by single bonds. $CH_3—CH_2—O—CH_2—CH_2—CH_3$
b. Ethyl cyclopropyl ether has a two-carbon group and a three-carbon cyclo alkyl group attached to oxygen by single bonds.

$CH_3—CH_2—O—$(cyclopropyl)

c. Methoxycyclopentane has a one-carbon group and a five-carbon cycloalkyl group attached to oxygen by single bonds.

OCH_3 on cyclopentane

d. 1-ethoxy-2-methylbutane has a four-carbon chain with a methyl attached to carbon 2 and an ethoxy attached to carbon 1.

```
                      CH3
                       |
CH3—CH2-O-CH2—CH—CH2—CH3
```

e. 2,3-dimethoxypentane has a five-carbon chain with two methoxy groups attached; one to carbon 2 and the other to carbon 3.

```
        OCH3
         |
CH3—CH—CH—CH2—CH3
         |
        OCH3
```

14.19 **a.** Constitutional isomers ($C_5H_{12}O$) have the same formula, but different arrangements.
b. Different compounds have different molecular formulas.
c. Constitutional isomers ($C_5H_{12}O$) have the same formula, but different arrangements.

14.21 The heterocyclic ethers with five atoms including one oxygen are named *furan*; six atoms including one oxygen are *pyrans*. A six-atom cyclic ether with two oxygen atoms is *dioxane*.

a. tetrahydrofuran **b.** 3-methylfuran **c.** 5-methyl-1, 3-dioxane

14.23 **a.** methanol; hydrogen bonding of alcohols gives higher boiling points than alkanes.
b. 1-butanol; alcohols hydrogen bond, but ethers cannot.
c. 1-butanol; hydrogen bonding of alcohols gives higher boiling points than alkanes.

14.25 **a.** yes, alcohols with 1-4 carbon atoms hydrogen bond with water
b. yes; the water can hydrogen bond to the O in ether
c. no; a carbon chain longer than 4 carbon atoms diminishes the effect of the —OH group.
d. no; alkanes are nonpolar and do not hydrogen bond
e. yes; the —OH in phenol ionizes in water, which makes it soluble.

14.27 Dehydration is the removal of an —OH and a —H from adjacent carbon atoms.

a. $CH_3—CH_2—CH{=}CH_2$ **b.**

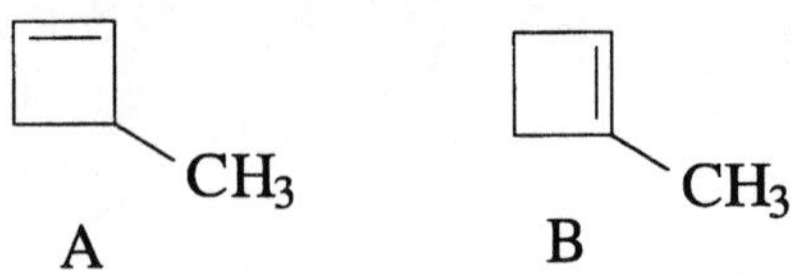

c. In c, there are two possible products A and B. B will be the major product, since the hydrogen is removed from the carbon that has the smaller number of hydrogens.

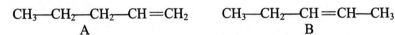

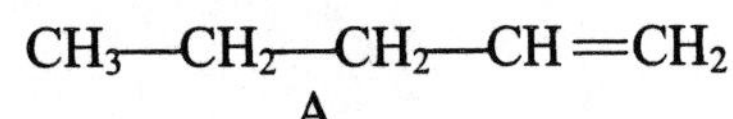

d. In d, there are two possible products A and B. B will be the major product, since the hydrogen is removed from the carbon that has the smaller number of hydrogen atoms attached.

$CH_3—CH_2—CH_2—CH{=}CH_2$ (A) $CH_3—CH_2—CH{=}CH—CH_3$ (B)

14.29 An ether is formed when H_2O is eliminated from two alcohols; the alkyl portion of one alcohol combines with the alkoxy portion of the other alcohol.

a. $CH_3—O—CH_3$ **b.** $CH_3—CH_2—CH_2—O—CH_2—CH_2—CH_3$

14.31 Alcohols can produce alkenes and ethers by the loss of water (dehydration).

a. $CH_3—CH_2—OH$
b. Since this ether has two different alkyl groups, it must be formed from two alcohols.
$CH_3—OH + CH_3—CH_2—OH$

c. OH

14.33 **a.** A primary alcohol oxidizes to an aldehyde and then to a carboxylic acid.

$CH_3—CH_2—CH_2—CH_2—\overset{O}{\overset{\|}{C}}—H$ then $CH_3—CH_2—CH_2—CH_2—\overset{O}{\overset{\|}{C}}—OH$

b. A secondary alcohol oxidizes to a ketone. $CH_3—CH_2—\overset{O}{\overset{\|}{C}}—CH_3$

c. A secondary alcohol oxidizes to a ketone.

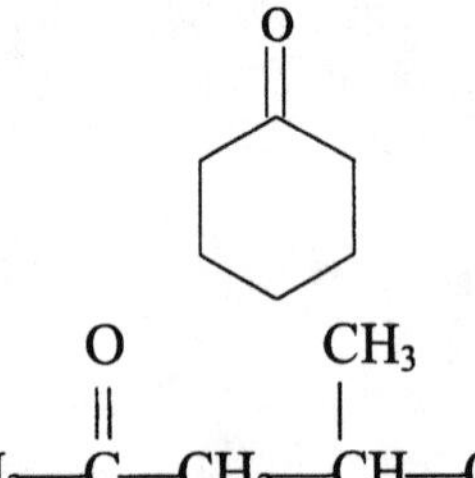

d. A secondary alcohol oxidizes to a ketone. $CH_3—\overset{O}{\overset{\|}{C}}—CH_2—\overset{CH_3}{\overset{|}{CH}}—CH_3$

e. A primary alcohol oxidizes to an aldehyde and then to a carboxylic acid.

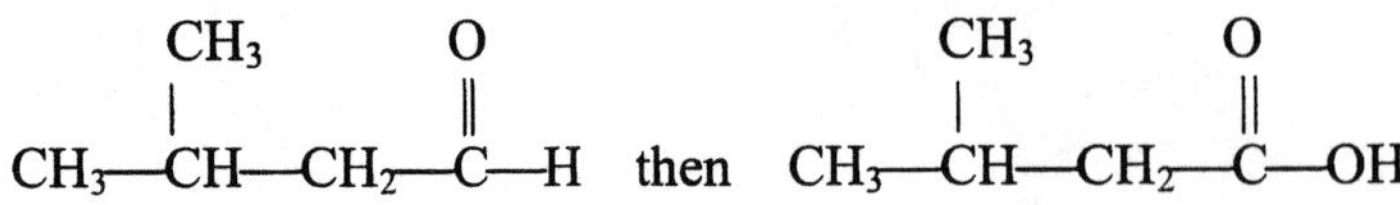

14.35 **a.** An aldehyde is the product of the oxidation of a primary alcohol CH_3—OH.

b. A ketone is the product of the oxidation of a secondary alcohol.

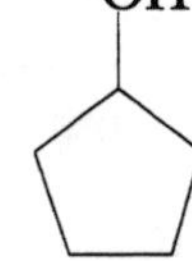

c. A ketone is the product of the oxidation of a secondary alcohol. $CH_3—CH(OH)—CH_2—CH_3$

d. An aldehyde is the product of the oxidation of a primary alcohol. CH_2OH

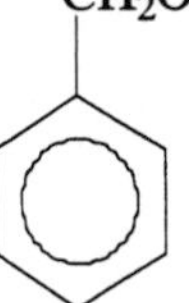

e. A ketone is the product of the oxidation of a secondary alcohol.

OH

CH_3

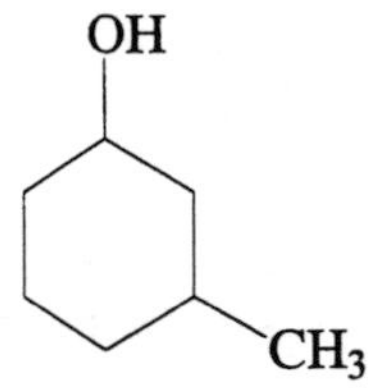

14.37 **a.** 2° **b.** 1° **c.** 1° **d.** 2° **e.** 1° **f.** 3°

14.39

a. alcohol **b.** ether **c.** thiol **d.** alcohol

e. ether **f.** cyclic ether **g.** alcohol **h.** phenol

14.41

a. 2-chloro-4-methylcyclohexanol

b. methyoxy benzene; methyl phenyl ether

c. 2-propanethiol

d. 2,4-dimethyl-2-pentanol

e. 1-methoxypropane; methyl propyl ether

f. 3-methyl furan

g. 4-bromo-2-pentanol

h. *meta*-cresol

14.43

a. 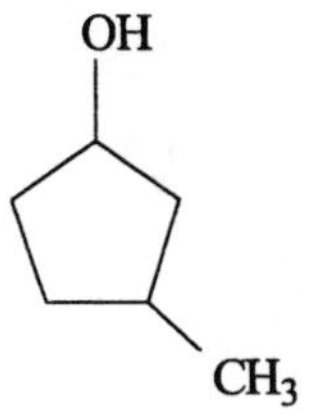 OH, CH_3

b. 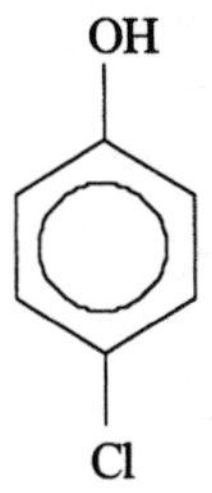 OH, Cl

c. $H_3C—CH(CH_3)—CH(OH)—CH_2—CH_3$

d. 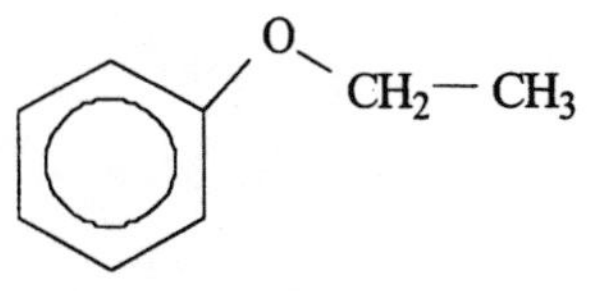 O, $CH_2—CH_3$

e. $CH_3—CH_2—CH(SH)—CH_2—CH_3$

f. 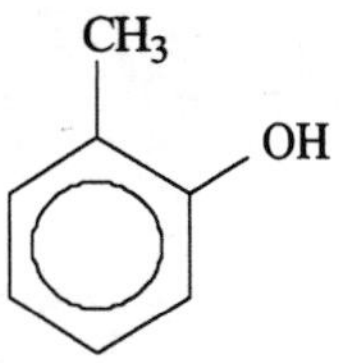 CH_3, OH

g. 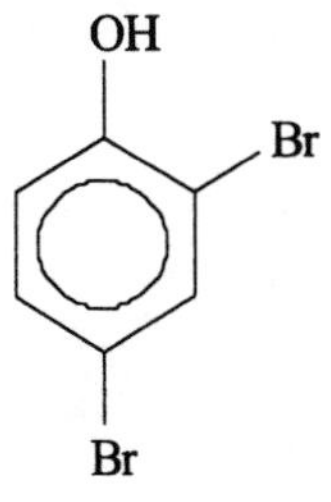 OH, Br, Br

14.45 **a.** Glycerol is used in skin lotions.
b. 1,2-ethanediol; ethylene glycol is used in antifreeze
c. Ethanol is produced by fermentation of grains and sugars.

14.47 Write the carbon chain first, and place the —OH on the carbon atoms in the chain to give different structural formulas. Shorten the chain by one carbon, and attach a methyl group and —OH group to give different compounds.

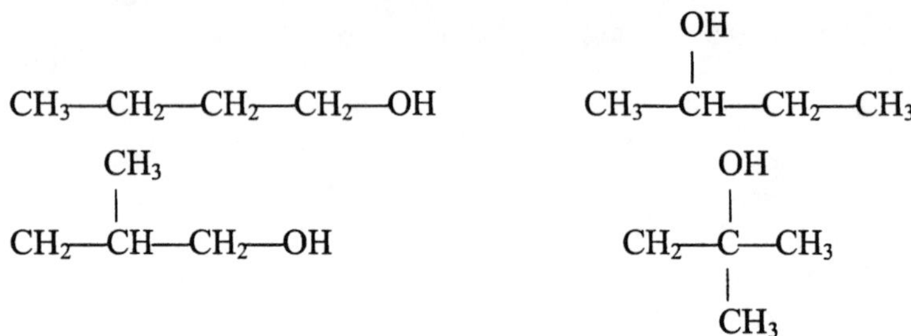

14.49 **a.** 1-propanol; hydrogen bonding **b.** 1-propanol; hydrogen bonding
c. 1-butanol; larger molar mass

14.51 **a.** soluble; hydrogen bonding
b. soluble; hydrogen bonding
c. insoluble; carbon chains over four carbon atoms diminishes effect of polar —OH on hydrogen bonding

14.53 **a.** $CH_3—CH{=}CH_2$ **b.** $CH_3—CH_2—\overset{\overset{\displaystyle O}{\|}}{C}—H$

c. $CH_3—CH{=}CH—CH_3$ **d.** $CH_3—CH_2—\overset{\overset{\displaystyle O}{\|}}{C}—CH_3$

e. $CH_3—CH_2—CH_2—O—CH_2—CH_2—CH_3$

f.

g. 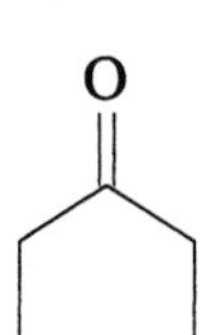

14.55 **a.** $CH_3—CH_2—CH_2—OH \xrightarrow{H^+,\ heat} CH_2—CH{=}CH_2 + HCl \longrightarrow CH_3—\underset{}{\overset{\overset{\displaystyle Cl}{|}}{C}H}—CH_3$

b. $CH_3—\overset{\overset{\displaystyle OH}{|}}{\underset{\underset{\displaystyle CH_3}{|}}{C}}—CH_3 \xrightarrow{H^+,\ heat} CH_3—\underset{\underset{\displaystyle CH_3}{|}}{C}{=}CH_2 + H_2 \xrightarrow{Pt} CH_3—\underset{\underset{\displaystyle CH_3}{|}}{C}H—CH_3$

c. 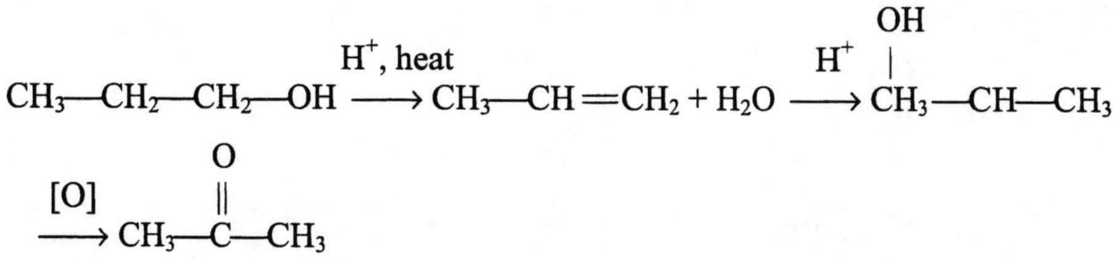 $CH_3{-}CH_2{-}CH_2{-}OH \xrightarrow{H^+,\ heat} CH_3{-}CH{=}CH_2 + H_2O \xrightarrow{H^+} CH_3{-}\underset{}{\overset{OH}{\overset{|}{C}H}}{-}CH_3$

$\xrightarrow{[O]} CH_3{-}\overset{O}{\overset{\|}{C}}{-}CH_3$

14.57 Testosterone contains cycloalkene, alcohol and ketone functional groups.

14.59 4-hexyl-1, 3-benzenediol tells us that there is a six carbon group attached to carbon 4 of a benzene ring and hydroxyls attached to carbons 1 and 3.

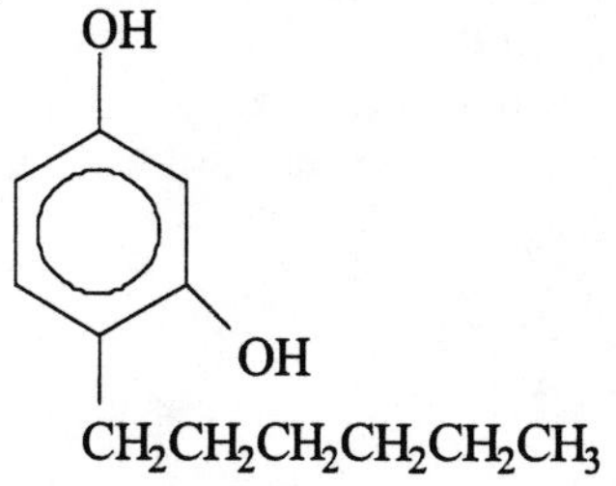

OH

OH

$CH_2CH_2CH_2CH_2CH_2CH_3$

14.61 **a.** 2,5-dichlorophenol is a benzene ring with a hydroxyl on carbon 1 and chlorine atoms on carbons 2 and 5.

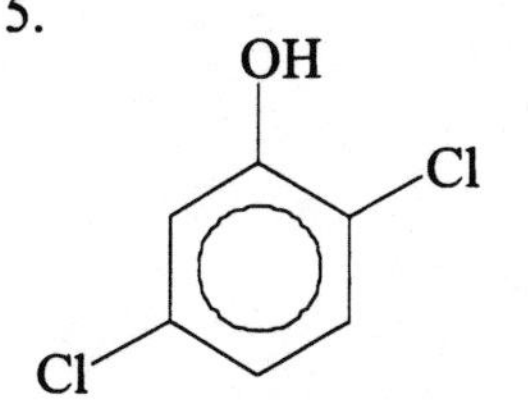

OH

Cl

Cl

b. 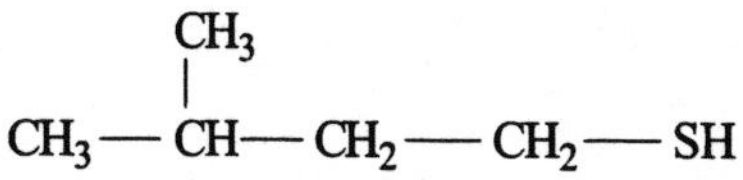 $CH_3{-}\overset{CH_3}{\overset{|}{C}H}{-}CH_2{-}CH_2{-}SH$

CH_3, H, C=C, H, $CH_2{-}SH$

c. OH, Cl, Cl, Cl, Cl, Cl

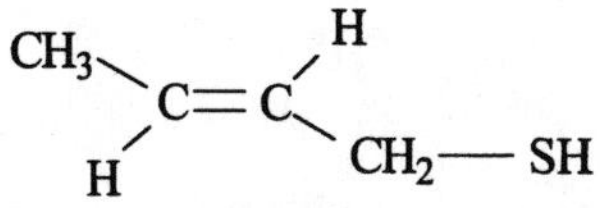

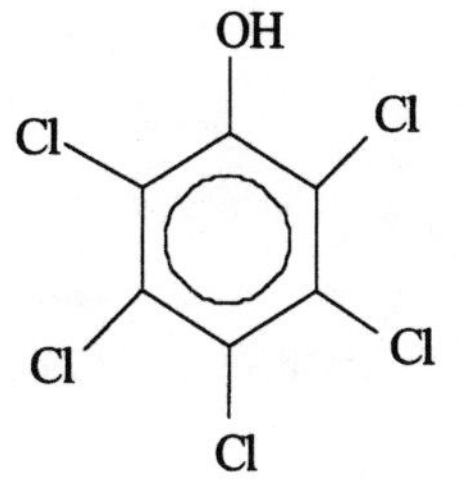

15

Aldehydes, Ketones, and Chiral Molecules

Study Goals

- Name and write the condensed structural formulas for aldehydes and ketones.
- Describe some important aldehydes and ketones.
- Identify the chiral carbon atoms in organic molecules.
- Write equations for the oxidation of aldehydes and for the reduction of aldehydes and ketones.
- Draw the structural formulas of hemiacetals and acetals produced from the addition of alcohols to aldehydes and ketones.

Think About It

1. What are the functional groups of an aldehyde and ketone?

2. How is an alcohol changed to an aldehyde or ketone?

3. When are mirror images not superimposable?

4. How are hemiacetals and acetals formed?

Key Terms

Match the following terms with the statements shown below.

a. chiral carbon **b.** hemiacetal **c.** Fischer projection **d.** aldehyde **e.** ketone

1. An organic compound with a carbonyl group attached to two alkyl groups

2. A carbon that is bonded to four different groups

3. The product that forms when an alcohol adds to an aldehyde or a ketone

4. A system for drawing chiral molecules that uses horizontal lines for bonds coming forward, and vertical lines for bonds going back with the chiral atom at the center

5. An organic compound that contains a carbonyl group and a hydrogen atom at the end of the carbon chain

Answers **1.** e **2.** a **3.** b **4.** c **5.** d

15.1 Structure and Bonding

- In an aldehyde, the carbonyl group appears at the end of a carbon chain attached to at least one hydrogen atom.
- In a ketone, the carbonyl group occurs between carbon groups and has no hydrogens attached to it.

◆ **Learning Exercise 15.1A**

Classify each of the following compounds.

A. alcohol **B.** aldehyde **C.** ketone **D.** ether **E.** thiol

____ **1.** $CH_3—CH_2—CH_2—\overset{\overset{\displaystyle O}{\|}}{C}—H$ ____ **2.** $CH_3—CH_2—CH_2—OH$

____ **3.** $CH_3—CH_2—\overset{\overset{\displaystyle O}{\|}}{C}—CH_2—CH_3$ ____ **4.** $CH_3—CH_2—O—CH_3$

____ **5.** $CH_3—\overset{\overset{\displaystyle O}{\|}}{C}—CH_2—CH_3$ ____ **6.** $CH_3—\overset{\overset{\displaystyle O}{\|}}{C}—H$

____ **7.** $CH_3—CH_2—\overset{\overset{\displaystyle SH}{|}}{C}H—CH_3$ ____ **8.** $CH_3—CH_2—\overset{\overset{\displaystyle OH}{|}}{C}H—CH_3$

Answers **1.** B **2.** A **3.** C **4.** D
5. C **6.** B **7.** E **8.** A

◆ **Learning Exercise 15.1B**

Indicate if each of the following pairs of compounds are constitutional isomers (C), the same compound (S), or different compounds (D).

1. ____ $CH_3—CH_2—\overset{\overset{\displaystyle O}{\|}}{C}—H$ and $CH_3—\overset{\overset{\displaystyle O}{\|}}{C}—CH_3$

2. ____ [skeletal structure of a ketone] and [skeletal structure of a ketone]

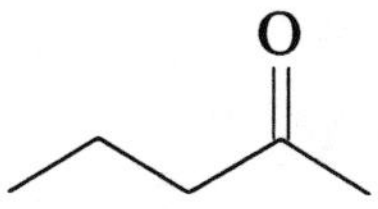

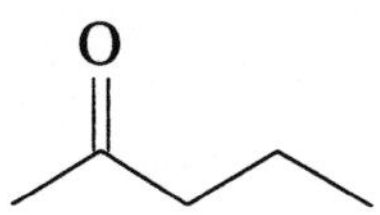

3. ____ $CH_3—\overset{\overset{\displaystyle O}{\|}}{C}—CH_2—CH_3$ and $CH_3—\overset{\overset{\displaystyle O}{\|}}{C}—CH_2—CH_2—CH_3$

Answers **1.** C **2.** S **3.** D

15.2 Naming Aldehydes and Ketones

In the IUPAC system, aldehydes and ketones are named by replacing the *e* in the longest chain containing the carbonyl group with *al* for aldehydes, and *one* for ketones. The location of the carbonyl group in a ketone is given if there are more than four carbon atoms in the chain.

$CH_3—\overset{\overset{\displaystyle O}{\|}}{C}—H$
ethanal
(acetaldehyde)

$CH_3—\overset{\overset{\displaystyle O}{\|}}{C}—CH_3$
propanone
(dimethyl ketone)

◆ Learning Exercise 15.2A

Write the correct IUPAC (or common name) for the following aldehydes.

1. 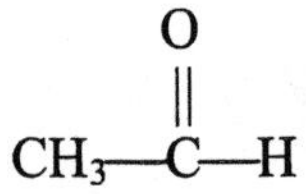 $CH_3—\overset{\overset{\displaystyle O}{||}}{C}—H$

2. $CH_3—CH_2—CH_2—CH_2—\overset{\overset{\displaystyle O}{||}}{C}—H$

3. $CH_3—CH_2—\overset{\overset{\displaystyle CH_3}{|}}{CH}—CH_2—CH_2—\overset{\overset{\displaystyle O}{||}}{C}—H$

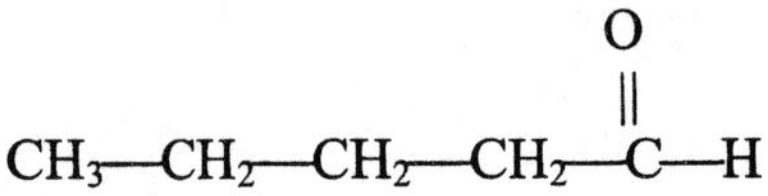

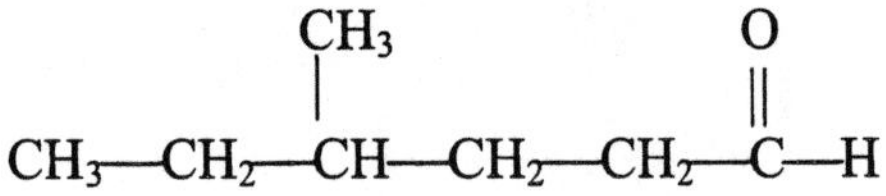

4. 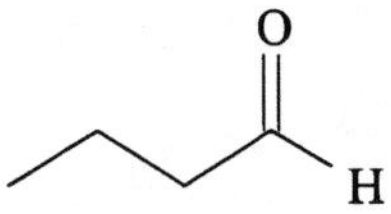 (skeletal structure: O, H)

5. 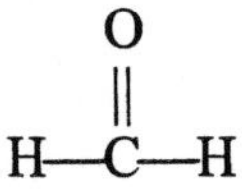 $H—\overset{\overset{\displaystyle O}{||}}{C}—H$

Answers
1. ethanal; acetaldehyde
2. pentanal
3. 4-methylhexanal
4. butanal; butyraldehyde
5. methanal; formaldehyde

◆ Learning Exercise 15.2B

Write the IUPAC (or common name) for the following ketones.

1. 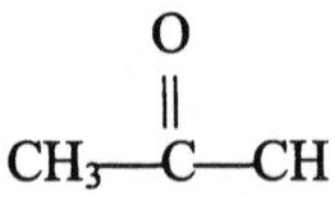 $CH_3—\overset{\overset{\displaystyle O}{||}}{C}—CH$

2. 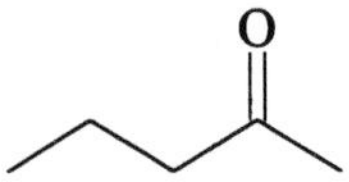 (skeletal structure: O)

3. $CH_3—CH_2—\overset{\overset{\displaystyle O}{||}}{C}—CH_2—CH_3$

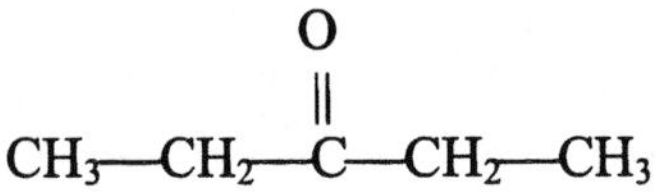

4. 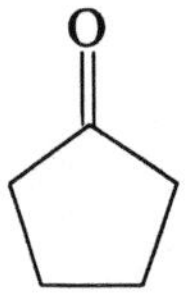 (cyclopentane ring with =O)

5. 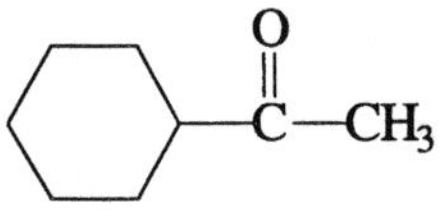 (cyclohexane ring)—$\overset{\overset{\displaystyle O}{||}}{C}—CH_3$

Answers
1. propanone; dimethyl ketone, acetone
2. 2-pentanone; methyl propyl ketone
3. 3-pentanone; diethyl ketone
4. cyclopentanone
5. cyclohexyl methyl ketone

◆ Learning Exercise 15.2C

Write the correct condensed formulas for the following.

1. ethanal
2. α-methylbutyraldehyde
3. 2-chloropropanal
4. ethylmethylketone
5. 3-hexanone
6. benzaldehyde

Answers

1. $CH_3—C(=O)—H$
2. $CH_3—CH_2—CH(CH_3)—C(=O)—H$
3. $CH_3—CH(Cl)—C(=O)—H$
4. $CH_3—CH_2—C(=O)—CH_3$
5. $CH_3—CH_2—C(=O)—CH_2—CH_2—CH_3$
6. $C_6H_5—C(=O)H$ (benzene ring—CH=O)

15.3 Some Important Aldehydes and Ketones

Match each of the following important aldehydes and ketones with a characteristic or use of that compound.

a. acetone **b.** formaldehyde **c.** butanedione **d.** benzaldehyde

1. ____ 40% solution is used to preserve biological specimens
2. ____ provides the flavor of butter in margarine
3. ____ a solvent used to remove paint and nail polish
4. ____ provides the odor and flavor of almonds
5. ____ produced in uncontrolled diabetes or fasting

Answers **1.** b **2.** c **3.** a **4.** d **5.** a

15.4 Physical Properties

- The polarity of the carbonyl group makes aldehydes and ketones of one to four carbon atoms soluble in water.

◆ Learning Exercise 15.4A

Indicate the compound with the highest boiling point in each of the following groups of compounds.

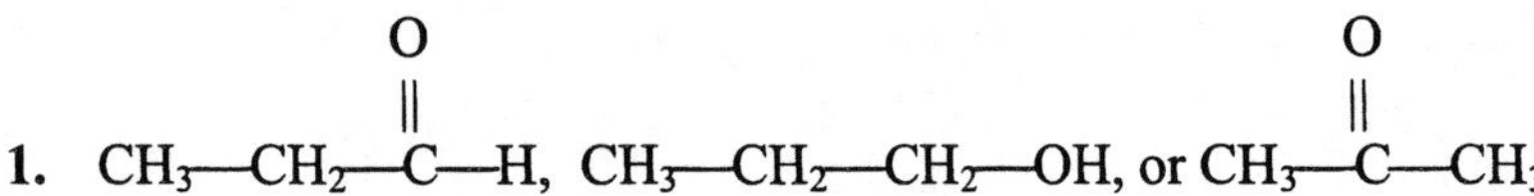

1. $CH_3—CH_2—C(=O)—H$, $CH_3—CH_2—CH_2—OH$, or $CH_3—C(=O)—CH_3$
2. acetaldehyde or propionaldehyde
3. propanone or butanone
4. methylcyclohexane or cyclohexanone

Answers 1. $CH_3—CH_2—CH_2—OH$ 2. propionaldehyde 3. butanone 4. cyclohexanone

◆ Learning Exercise 15.4B

Indicate whether each of the following compounds is soluble (S) or not soluble (NS) in water.

1. _____ 3-hexanone 2. _____ propanal 3. _____ acetaldehyde
4. _____ butanal 5. _____ cyclohexanone

Answers 1. NS 2. S 3. S 4. S 5. NS

15.5 Chiral Molecules

- Chiral molecules have mirror images that cannot be superimposed.
- In a chiral molecule, there is one or more carbon atoms attached to four different atoms or groups.
- The mirror images of a chiral molecule represent two different molecules called enantiomers.
- In a Fischer projection (straight chain), the prefixes D- and L- are used to distinguish between the mirror images. In D-glyceraldehyde, the —OH is on the right of the chiral carbon; it is on the left in L-glyceraldehyde.

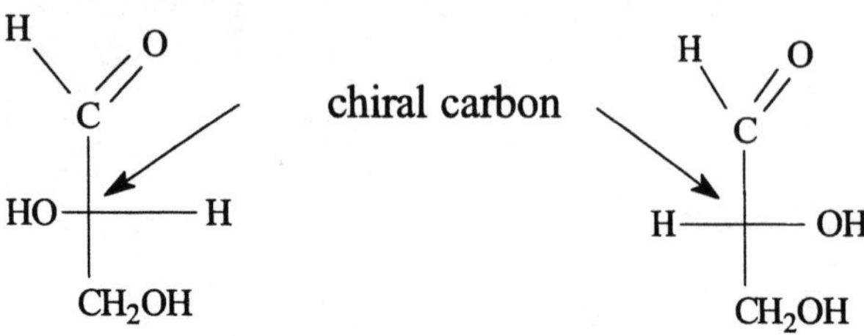

L-Glyceraldehyde D-Glyceraldehyde

◆ Learning Exercise 15.5A

Indicate whether the following objects would be chiral or not.

1. a piece of plain computer paper ____
2. weight-lifting glove ____
3. a baseball cap ____
4. a volleyball net ____
5. your left foot ____

Answers **1.** not chiral **2.** chiral **3.** not chiral **4.** not chiral **5.** chiral

◆ Learning Exercise 15.5B

State whether each of the following molecules is chiral or not chiral.

1. Cl above; H—C—Cl; CH_3 below ________

2. 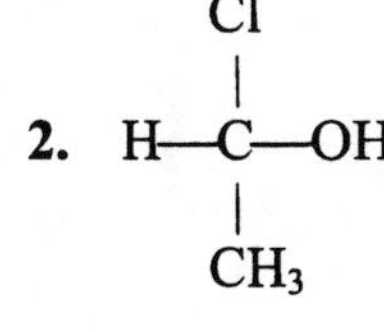

____

3. 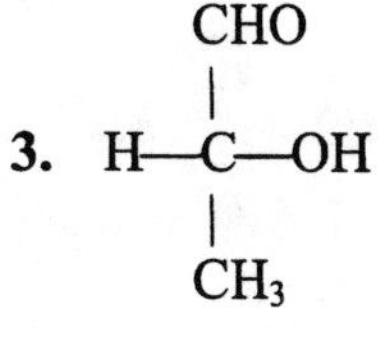

____

Answers **1.** not chiral **2.** chiral **3.** chiral

◆ Learning Exercise 15.5C

Identify the following as a feature that is characteristic of a chiral compound or not.

1. the central atom is attached to two identical groups ________
2. contains a carbon attached to four different groups ________
3. has identical mirror images ________

Answers **1.** not chiral **2.** chiral **3.** chiral

◆ Learning Exercise 15.5D

Indicate whether each pair of Fischer projections represents enantiomers (E) or identical structures (I).

a. COOH above; HO—+—H; CH_2OH below **and** COOH above; H—+—OH; CH_2OH below

b. 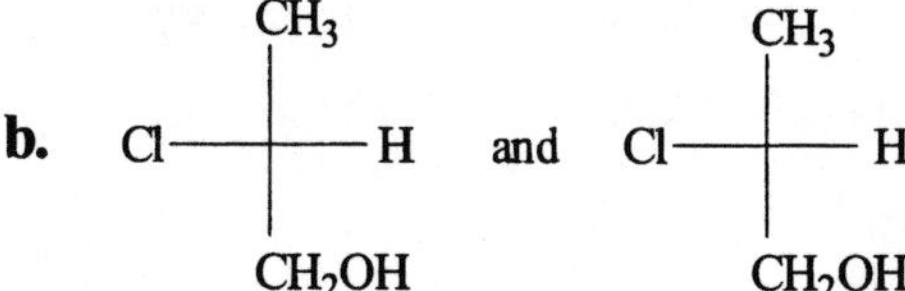

c. 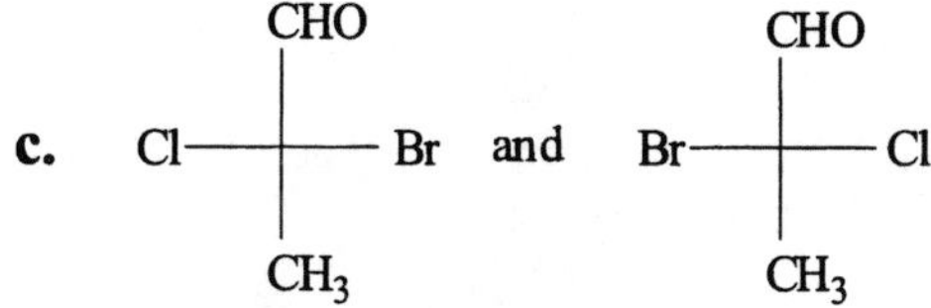

d. COOH above; H—+—OH; COOH below **and** COOH above; HO—+—H; COOH below

Answers **1.** E **2.** I **3.** E **4.** I

15.6 Oxidation and Reduction

- Using an oxidizing agent, primary alcohols oxidize to aldehydes, which usually oxidize further to carboxylic acids. Secondary alcohols are oxidized to ketones, but tertiary alcohols do not oxidize.

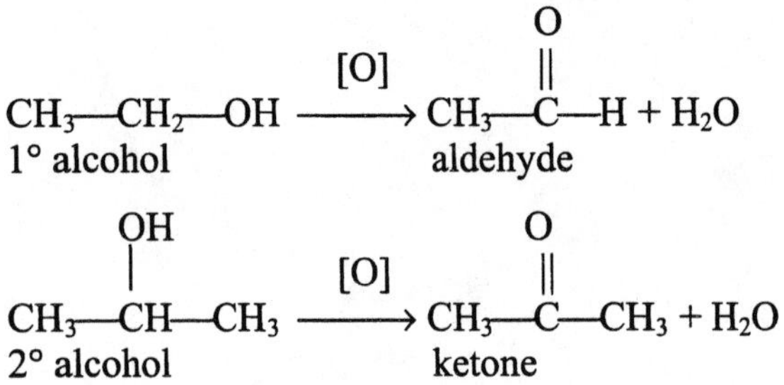

$$\underset{1°\ \text{alcohol}}{CH_3{-}CH_2{-}OH} \xrightarrow{[O]} \underset{\text{aldehyde}}{CH_3{-}\overset{O}{\overset{\|}{C}}{-}H} + H_2O$$

$$\underset{2°\ \text{alcohol}}{CH_3{-}\overset{OH}{\overset{|}{C}H}{-}CH_3} \xrightarrow{[O]} \underset{\text{ketone}}{CH_3{-}\overset{O}{\overset{\|}{C}}{-}CH_3} + H_2O$$

- Aldehydes and ketones are reduced when hydrogen is added in the presence of a metal catalyst to produce primary or secondary alcohols.

◆ Learning Exercise 15.6A

Write the structural formula of the alcohol that oxidized to give each of the following compounds.

1. 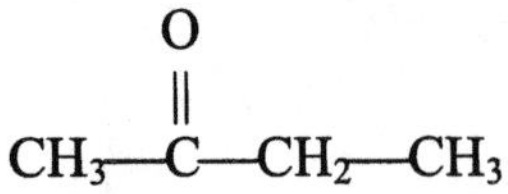 $CH_3{-}\overset{O}{\overset{\|}{C}}{-}CH_2{-}CH_3$

2. (cyclohexanone: six-membered ring with C=O)

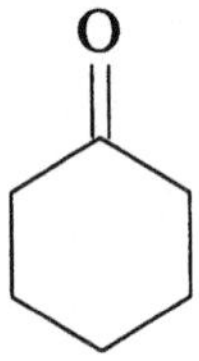

3. $CH_3{-}CH_2{-}\overset{O}{\overset{\|}{C}}{-}H$

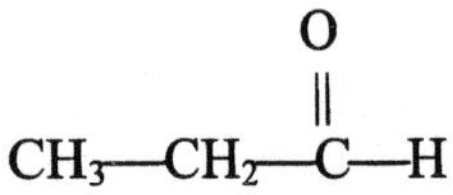

Answers 1. 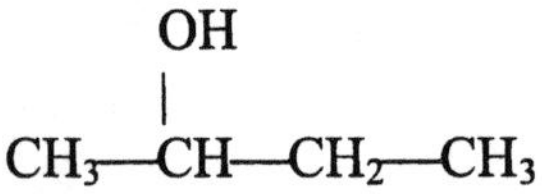 $CH_3{-}\overset{OH}{\overset{|}{C}H}{-}CH_2{-}CH_3$ 2. (cyclohexanol: six-membered ring with OH) 3. $CH_3{-}CH_2{-}CH_2{-}OH$

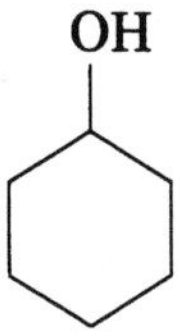

◆ Learning Exercise 15.6B

Indicate the compound in each of the following pairs that will oxidize.

1. propanal or propanone ____________________
2. butane or butanal ____________________
3. ethane or acetaldehyde ____________________

Answers 1. propanal 2. butanal 3. acetaldehyde

◆ Learning Exercise 15.6C

Write the reduction products for the following.

1. $CH_3-\overset{\overset{\large O}{\|}}{C}-CH_3 + H_2 \xrightarrow{Pt}$

2. $CH_3-CH_2-\overset{\overset{\large O}{\|}}{C}-H + H_2 \xrightarrow{Pt}$

3. $CH_3-\overset{\overset{\large O}{\|}}{C}-H + H_2 \xrightarrow{Pt}$

4. 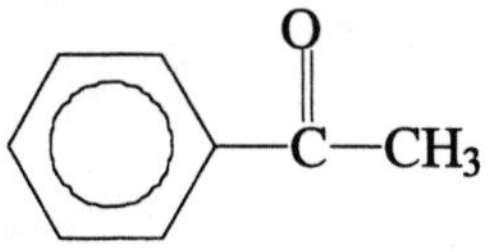 $C_6H_5-\overset{\overset{\large O}{\|}}{C}-CH_3 + H_2 \xrightarrow{Pt}$ (benzene ring attached to C)

5. 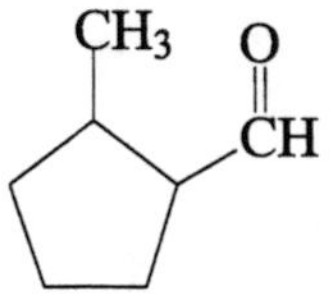 (cyclopentane ring bearing CH_3 and, on the adjacent carbon, $\overset{\overset{\large O}{\|}}{C}H$) $+ H_2 \xrightarrow{Pt}$

Answers 1. $CH_3-\overset{\overset{\large OH}{|}}{CH}-CH_3$ 2. $CH_3-CH_2-CH_2-OH$ 3. CH_3-CH_2-OH

4. 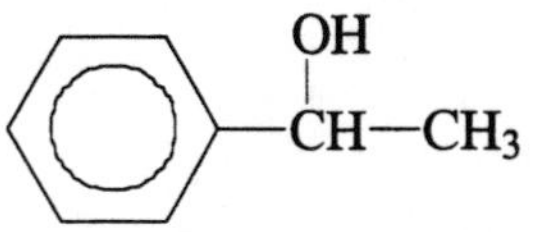 $C_6H_5-\overset{\overset{\large OH}{|}}{CH}-CH_3$ (benzene ring attached to CH) 5. (cyclopentane ring bearing CH_3 and, on the adjacent carbon, CH_2OH)

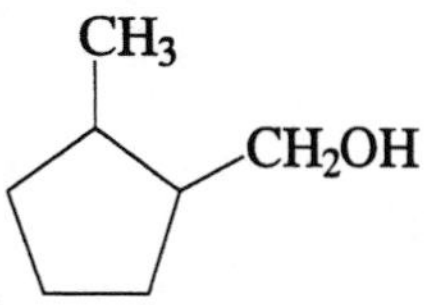

15.7 Addition Reactions

- Alcohols add to the carbonyl group of aldehyde and ketones.
- Hemiacetals form when one alcohol adds to aldehydes or ketones.
- Acetals form when a second alcohol molecule adds to hemiacetals.

◆ Learning Exercise 15.7A

Match the statements shown below with the following types of compounds.

A. hemiacetal **B.** acetal

1. ____ The product from the addition of one alcohol to an aldehyde
2. ____ The product from the addition of one alcohol to a ketone
3. ____ A compound that contains two ether groups
4. ____ A compound that consists of one ether group, an alcohol group, and two alkyl groups

Answers **1.** A **2.** A **3.** B **4.** A

◆ Learning Exercise 15.7B

Identify each of the following structural formulas as a hemiacetal, acetal, or neither.

1. $CH_3—O—CH_2—OH$

2.
```
      OH
      |
CH3—C—H
      |
      O—CH3
```

3.
```
      O—CH2—CH3
      |
CH3—C—CH3
      |
      O—CH2—CH3
```

4.
```
      O—CH3
      |
CH3—C—H
      |
      O—CH3
```

Answers **1.** hemiacetal **2.** hemiacetal **3.** acetal **4.** acetal

◆ Learning Exercise 15.7C

Write the structural formula of the hemiacetal and acetal products when methanol adds to propanone.

Answers

```
      OH                      O—CH3
      |                       |
CH3—C—O—CH3             CH3—C—O—CH3
      |                       |
      CH3                     CH3
   Hemiacetal                acetal
```

Check List for Chapter 15

You are ready to take the practice test for Chapter 15. Be sure that you have accomplished the following learning goals for this chapter. If you are not sure, review the section listed at the end of the goal. Then apply your new skills and understanding to the practice test.

After studying Chapter 15, I can successfully:

_____ Identify structural formulas as aldehydes and ketones (15.1).

_____ Give the IUPAC and common names of an aldehyde or ketone; draw the condensed structural formula from the name (15.2).

_____ Identify the uses of some common aldehydes and ketones (15.3).

_____ Compare the physical properties of aldehydes and ketones with alcohols and alkanes (15.4).

_____ Identify a molecule as a chiral or not: write the D- and L-Fischer projections (15.5),

_____ Write the structural formulas for reactants and products of the oxidation of alcohols or reduction of aldehydes and ketones (15.6).

_____ Write the structural formulas of the hemiacetals and acetals that form when alcohols add to aldehyde or ketones (15.7).

Practice Test for Chapter 15

Match the following compounds with the names given.

A. dimethyl ether **B.** acetaldehyde **C.** methanal **D.** dimethyl ketone **E.** propanal

1. ____ 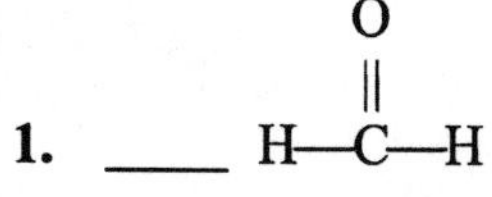 H—C(=O)—H

2. ____ CH_3—O—CH_3

3. ____ 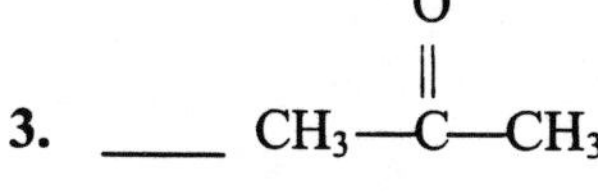 CH_3—C(=O)—CH_3

4. ____ CH_3—C(=O)—H

5. CH_3—CH_2—C(=O)—H

6. The compound with the higher boiling point is

A. CH_3—CH_2—CH_2—CH_3

B. CH_3—CH_2—CH_2—OH

C. CH_3—C(=O)—CH_3

D. CH_3—CH_2—C(=O)—H

E. CH_3—CH_2—O—CH_3

Complete questions 7 through 11 by indicating one of the products (A–E) formed in each of the following reactions.

A. primary alcohol **B.** secondary alcohol **C.** aldehyde
D. ketone **E.** carboxylic acid

7. ____ Oxid ation of a primary alcohol

8. ____ Oxidation of a secondary alcohol

9. ____ Oxidation of an aldehyde

10. ____ Reduction of a ketone

11. ____ Reduction of an aldehyde

12. Benedict's reagent will oxidize

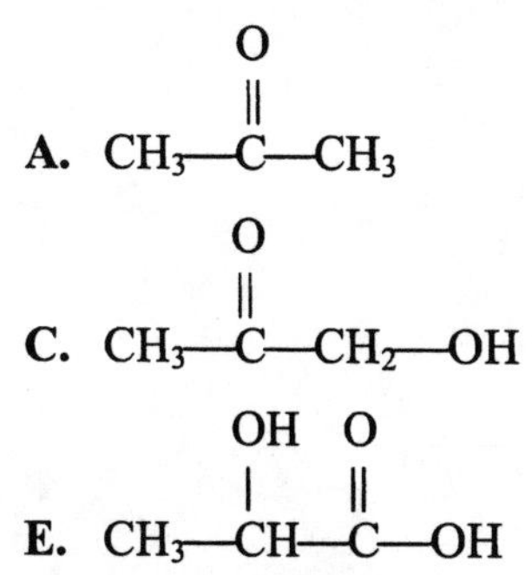

A. CH_3—C(=O)—CH_3

B. CH_3—CH(OH)—CH_2—OH

C. CH_3—C(=O)—CH_2—OH

D. CH_3—CH_2—C(=O)—H

E. CH_3—CH(OH)—C(=O)—OH

13. In the Tollen's test

A. an aldehyde is oxidized and Ag^+ reduced.
B. an aldehyde is reduced and Ag^+ oxidized.
C. a ketone is oxidized and Ag^+ reduced.
D. a ketone is reduced and Ag^+ oxidized.
E. all of these

Identify each of the following structural formulas as

14. alcohol **15.** ether **16.** hemiacetal **17.** acetal

A. $CH_3-CH_2-\underset{}{\overset{OH}{\overset{|}{C}}}H-CH_3$

B. 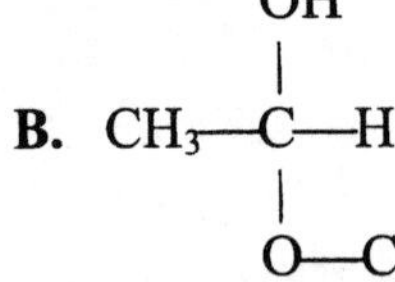

C. $CH_3-\overset{O-CH_2-CH_3}{\overset{|}{\underset{\underset{O-CH_2-CH_3}{|}}{C}}}-CH_3$

D. $CH_3-\overset{O-CH_3}{\overset{|}{\underset{\underset{CH_3}{|}}{C}}}-CH_3$

18. The structural formula for 4-bromo-3-methylcyclohexanone is

A. O, Br, CH_3

B. O, H_3C, Br

C. O, Br, CH_3

D. O, Br, CH_3

E. O, Br, CH_3

19. The name of this compound is

$$CH_3-\overset{O}{\overset{\|}{C}}-CH_2-\overset{CH_3}{\overset{|}{C}}H-CH_2-\overset{O}{\overset{\|}{C}}-H$$

A. 2-oxo-4-methylhexanal.
B. 3-methyl-2-oxohexanal.
C. 5-oxo-3-methylhexanal.
D. 1-aldo-4-methyl-2-hexanol.
E. 5-oxo-3-methyl-1-hexanone.

20. The reaction of an alcohol with an aldehyde is called a(n)

A. elimination.
B. addition.
C. substitution.
D. hydrolysis.
E. oxidation.

Identify each of the following as enantiomers (E), identical (I), or different (D) compounds.

21. $\begin{array}{c} COOH \\ HO-\!\!\!-\!\!\!+\!\!\!-\!\!\!-H \\ CH_2OH \end{array}$ and $\begin{array}{c} COOH \\ H-\!\!\!-\!\!\!+\!\!\!-\!\!\!-OH \\ CH_2OH \end{array}$

22. $\begin{array}{c} CH_3 \\ Cl-\!\!\!-\!\!\!+\!\!\!-\!\!\!-H \\ CH_2OH \end{array}$ and $\begin{array}{c} CH_3 \\ Cl-\!\!\!-\!\!\!+\!\!\!-\!\!\!-H \\ CH_2OH \end{array}$

23. $\begin{array}{c} CHO \\ Cl-\!\!\!-\!\!\!+\!\!\!-\!\!\!-Cl \\ CH_3 \end{array}$ and $\begin{array}{c} CHO \\ Cl-\!\!\!-\!\!\!+\!\!\!-\!\!\!-Cl \\ CH_3 \end{array}$

24. $\begin{array}{c} COOH \\ H-\!\!\!-\!\!\!+\!\!\!-\!\!\!-OH \\ COOH \end{array}$ and $\begin{array}{c} COOH \\ HO-\!\!\!-\!\!\!+\!\!\!-\!\!\!-H \\ COOH \end{array}$

25. $\begin{array}{c} CHO \\ Cl-\!\!\!-\!\!\!+\!\!\!-\!\!\!-Br \\ CH_3 \end{array}$ and $\begin{array}{c} COOH \\ Br-\!\!\!-\!\!\!+\!\!\!-\!\!\!-Cl \\ CH_3 \end{array}$

Answers to Practice Test

1. C	**2.** A	**3.** D	**4.** B	**5.** E
6. B	**7.** C, E	**8.** D	**9.** E	**10.** B
11. A	**12.** D	**13.** A	**14.** A	**15.** D
16. B	**17.** C	**18.** B	**19.** C	**20.** B
21. E	**22.** I	**23.** I	**24.** I	**25.** D

Answers and Solutions to Selected Text Problems

15.1 a. ketone **b.** aldehyde **c.** ketone **d.** aldehyde

15.3 a. 1 **b.** 1 **c.** 2

15.5 a. propanal **b.** 2-methyl-3-pentanone
c. 3-hydroxybutanal **d.** 2-pentanone
e. 3-methylcyclohexanone **f.** 4-chlorobenzaldehyde

15.7 a. acetaldehyde **b.** methyl propyl ketone
c. formaldehyde

15.9 a. $CH_3-\overset{\overset{O}{\|}}{C}-H$ **b.** $CH_3-\overset{\overset{O}{\|}}{C}-CH_2-\overset{\overset{OH}{|}}{CH}-CH_3$

c. $CH_3-\overset{\overset{Br}{|}}{CH}-\overset{\overset{Br}{|}}{CH}-\overset{\overset{O}{\|}}{C}-H$ **d.** $CH_3-\overset{\overset{O}{\|}}{C}-CH_2-CH_2-CH_2-CH_3$

e. $CH_3-CH_2-\overset{\overset{CH_3}{|}}{CH}-CH_2-\overset{\overset{O}{\|}}{C}-H$ **f.** $CH_3-\overset{\overset{O}{\|}}{C}-\overset{\overset{O}{\|}}{C}-H$

15.11

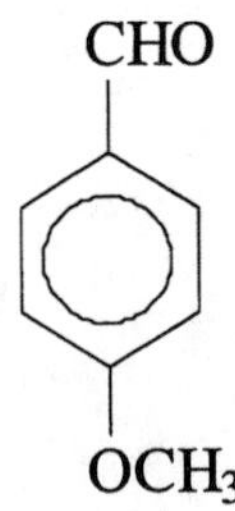

15.13 **a.** benzaldehyde **b.** acetone; propanone **c.** formaldehyde

15.15 **a.** $CH_3—CH_2—\overset{\overset{O}{\|}}{C}—H$ has a polar carbonyl group.
b. pentanal has more carbons and thus a higher molar mass.
c. 1-butanol can hydrogen bond with other 1-butanol molecules.

15.17 **a.** $CH_3—\overset{\overset{O}{\|}}{C}—\overset{\overset{O}{\|}}{C}—CH_3$: more hydrogen bonding
b. acetaldhyde; more hydrogen bonding
c. acetone; lower number of carbon atoms

15.19 No. The carbon chain diminishes the effect of the carbonyl group.

15.21 **a.** achiral **b.** chiral $CH_3—\underset{}{\overset{\overset{Br}{|}}{CH}}—CH_2CH_3$ (*chiral carbon*)

c. chiral $CH_3—\overset{\overset{Br}{|}}{CH}—\overset{\overset{O}{\|}}{C}—H$ (*chiral carbon*) **d.** achiral

15.23 **a.**

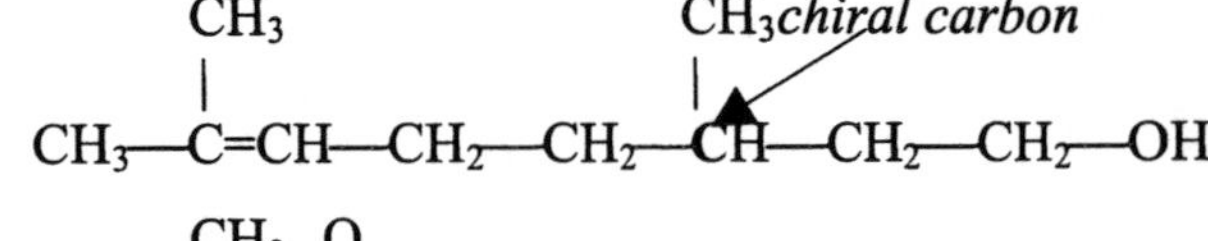

b. $H_2N—\overset{\overset{CH_3}{|}}{CH}—\overset{\overset{O}{\|}}{C}—OH$ (*chiral carbon*)

15.25

a. Fischer projection: H (top), HO (left), Br (right), CH_3 (bottom)

b. Fischer projection: CH_3 (top), Cl (left), Br (right), OH (bottom)

c. Fischer projection: CHO (top), HO (left), H (right), CH_2CH_3 (bottom)

15.27 **a.** identical **b.** enantiomers
c. identical **d.** enantiomers

15.29 **a.** An aldehyde is the product of the oxidation of a primary alcohol CH_3OH.

b. A ketone is the product of the oxidation of a secondary alcohol.

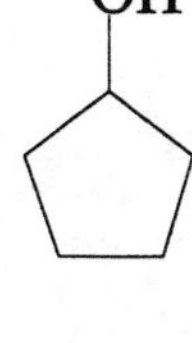

c. A ketone is the product of the oxidation of a secondary alcohol.

$CH_3—CH(OH)—CH_2—CH_3$

d. An aldehyde is the product of the oxidation of a primary alcohol. CH_2OH

e. A ketone is the product of the oxidation of a secondary alcohol.

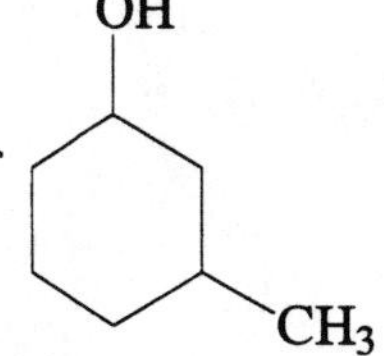

15.31 **a.** A primary alcohol will be oxidized to an aldehyde and then to a carboxylic acid.

$CH_3—CH_2—CH_2—CH_2—C(=O)—H$ then $CH_3—CH_2—CH_2—CH_2—C(=O)—OH$

b. A secondary alcohol will be oxidized to a ketone. $CH_3—CH_2—C(=O)—CH_3$

c. A secondary alcohol will be oxidized to a ketone.

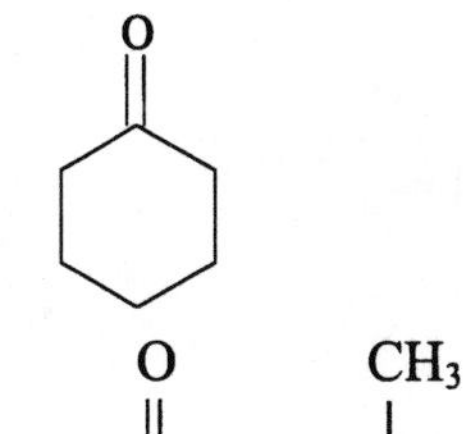

d. A secondary alcohol will be oxidized to a ketone. $CH_3—C(=O)—CH_2—CH(CH_3)—CH_3$

e. A primary alcohol will be oxidized to an aldehyde and then to a carboxylic acid.

$CH_3—CH(CH_3)—CH_2—C(=O)—H$ then $CH_3—CH(CH_3)—CH_2—C(=O)—OH$

15.33 In reduction, an aldehyde will give a primary alcohol and a ketone will give a secondary alcohol.

a. Butyraldehyde is the four carbon aldehyde; it will be reduced to a four-carbon, primary alcohol.

$CH_3—CH_2—CH_2—CH_2—OH$

b. Acetone is a three-carbon ketone; it will be reduced to a three-carbon secondary alcohol.

$$\begin{array}{c} \quad OH \\ \quad | \\ CH_3—CH—CH_3 \end{array}$$

c. 3-bromohexanal is a six-carbon aldehyde with bromine attached to carbon 3. It reduces to a six-carbon primary alcohol with bromine on carbon 3.

$$\begin{array}{c} \qquad\qquad\qquad Br \\ \qquad\qquad\qquad | \\ CH_3—CH_2—CH_2—CH—CH_2—CH_2\text{–}OH \end{array}$$

d. 2-methyl-3-pentanone is a five-carbon ketone with a methyl group attached to carbon 2. It will be reduced to a five-carbon secondary alcohol with a methyl attached to carbon 2.

$$\begin{array}{c} CH_3\ \ OH \\ |\quad\ \ | \\ CH_3—CH—CH\text{–}CH_2—CH_3 \end{array}$$

15.35 **a.**
$$\begin{array}{c} OH \\ | \\ CH_3—C—H \\ | \\ OH \end{array}$$
b.
$$\begin{array}{c} OH \\ | \\ H—C—H \\ | \\ OH \end{array}$$

15.37 **a.** hemiacetal **b.** hemiacetal **c.** acetal
d. hemiacetal **e.** acetal

15.39 A hemiacetal forms when an alcohol is added to the carbonyl of an aldehyde or ketone.

a.
$$\begin{array}{c} O—CH_3 \\ | \\ CH_3—C—H \\ | \\ OH \end{array}$$
b.
$$\begin{array}{c} O—CH_3 \\ | \\ CH_3—C—CH_3 \\ | \\ OH \end{array}$$

c. HO OCH$_3$ (cyclopentane ring carbon bearing HO and OCH_3)

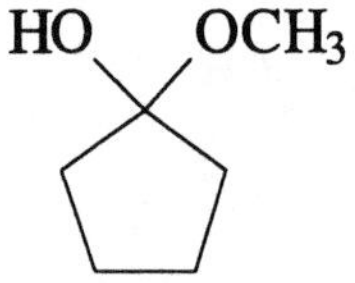

d.
$$\begin{array}{c} OCH_3 \\ | \\ CH_3—CH_2—CH_2—C—H \\ | \\ OH \end{array}$$

15.41 An acetal forms when a second molecule of alcohol reacts with a hemiacetal.

a.
$$\begin{array}{c} OCH_3 \\ | \\ CH_3—C—H \\ | \\ OCH_3 \end{array}$$
b.
$$\begin{array}{c} OCH_3 \\ | \\ CH_3—C—CH_3 \\ | \\ OCH_3 \end{array}$$

c. CH_3O OCH_3

d.
$$\begin{array}{c} \quad\quad\quad\quad\quad\quad\quad\quad CH_3 \\ \quad\quad\quad\quad\quad\quad\quad\quad | \\ CH_3{-}CH_2{-}CH_2{-}C{-}H \\ \quad\quad\quad\quad\quad\quad\quad\quad | \\ \quad\quad\quad\quad\quad\quad\quad\quad OCH_3 \end{array}$$

15.43 The carbonyl group consists of a sigma bond and a pi bond, which is an overlapping of the *p* orbitals of the carbon and oxygen atom.

15.45

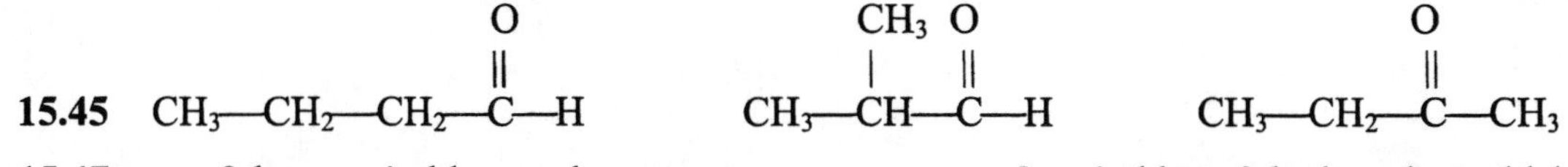

15.47
a. 2-bromo-4-chlorocyclopentanone
b. 4-chloro-3-hydroxybenzaldehyde
c. 3-chloropropanal; β-chloropropionaldehyde
d. 5-hydroxy-3-hexanone
e. 2-chloro-3-pentanone
f. 3-methylcyclohexanone

15.49
a. 3-methylcyclopentanone is a five-carbon cyclic structure with a methyl group located two carbons from the carbonyl group.

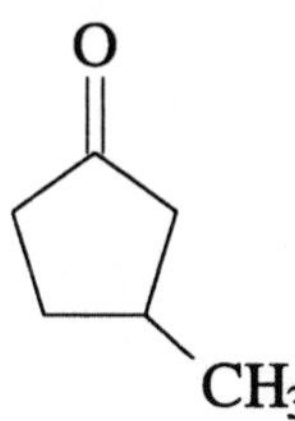

b. *p*-chlorobenzaldehyde is a benzene with an aldehyde group and a chlorine on carbon 4.

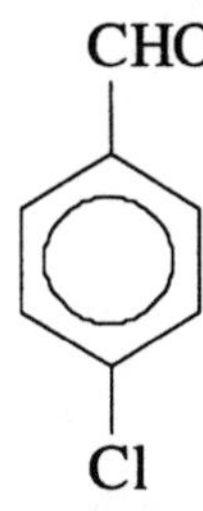

c. *β*-chloropropionaldehyde is a three carbon aldehyde with a chlorine located two carbons from the carbonyl.

$$\begin{array}{c} \quad\quad\quad\quad\quad\quad\quad O \\ \quad\quad\quad\quad\quad\quad\quad \| \\ Cl{-}CH_2{-}CH_2{-}C{-}H \end{array}$$

d. Butanone is a four-carbon ketone.

$$\begin{array}{c} O \\ \| \\ CH_3{-}C{-}CH_2{-}CH_3 \end{array}$$

e. This is a six-carbon aldehyde with a methyl group on carbon 3.

$$CH_3{-}CH_2{-}CH_2{-}\underset{}{\overset{CH_3}{\overset{|}{C}}}H{-}CH_2{-}\overset{O}{\overset{||}{C}}{-}H$$

f. This has a seven-carbon chain with a carbonyl group on carbon 2.

$$CH_3{-}\overset{O}{\overset{||}{C}}{-}CH_2{-}CH_2{-}CH_2{-}CH_2{-}CH_3$$

15.51 Compounds b, c, and d with oxygen atoms and less than five carbon atoms can hydrogen bond to be soluble.

15.53 **a.** $CH_3{-}CH_2{-}OH$; polar —OH group can hydrogen bond

b. $CH_3{-}CH_2{-}\overset{O}{\overset{||}{C}}{-}H$; polar carbonyl group can hydrogen bond

c. $CH_3{-}CH_2{-}CH_2{-}OH$; polar —OH group can hydrogen bond

15.55 A chiral carbon is bonded to four different groups.

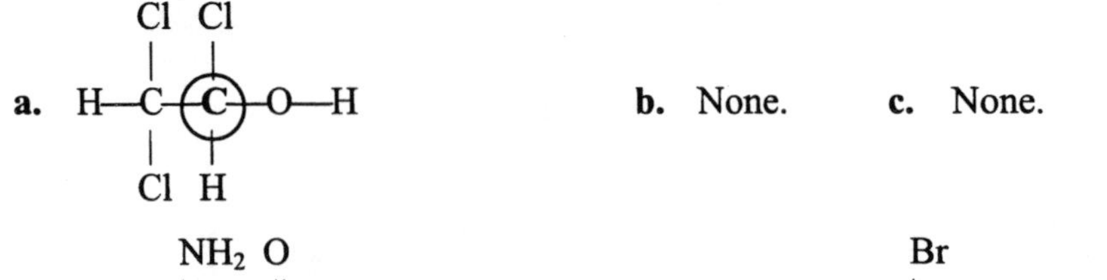

f. None.

15.57 Enantiomers are mirror images

a. identical
b. enantiomers
c. enantiomers (turn 180°)
d. enantiomers

15.59 Primary alcohols oxidize to aldehydes and then to carboxylic acids. Secondary alcohols oxidize to ketones.

a. $CH_3{-}CH_2{-}\overset{O}{\overset{||}{C}}{-}H \xrightarrow{\text{further oxidation}} CH_3{-}CH_2{-}\overset{O}{\overset{||}{C}}{-}OH$

b. $CH_3{-}\overset{O}{\overset{||}{C}}{-}CH_2{-}CH_2{-}CH_3$

c. $CH_3{-}CH_2{-}CH_2{-}\overset{O}{\overset{||}{C}}{-}OH$

d. (cyclohexanone structure: six-membered ring with C=O)

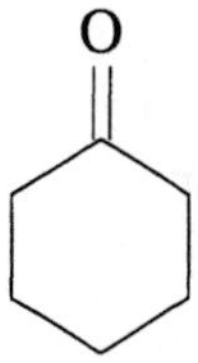

15.61 **a.** $CH_3-CH(OH)-CH_3$

b. $C_6H_5-CH_2-CH_2-OH$

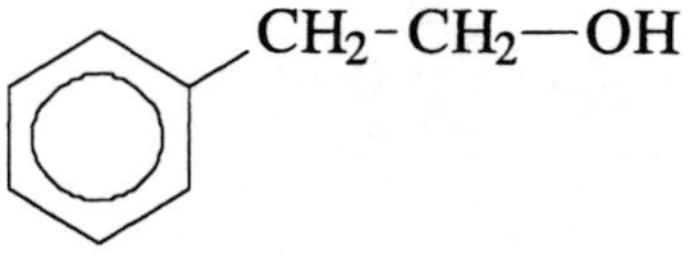

c. $CH_3-CH(CH_3)-CH_2-CH(OH)-CH_3$

15.63 **a.** $CH_3-CH{=}CH_2 + H_2O \xrightarrow{H^+} CH_3-CH(OH)-CH_3 \longrightarrow CH_3-C(=O)-CH_3$

propene → propanone

b. $CH_3-CH_2-CH_2-C(=O)-H + H_2 \xrightarrow{Ni} CH_3-CH_2-CH_2-CH_2-OH \xrightarrow{H^+,\ heat}$

$CH_3-CH_2-CH{=}CH_2 + Br_2 \longrightarrow CH_3-CH_2-CH(Br)-CH_2-Br$

c. $CH_3-CH_2-CH_2-C(=O)-H + H_2 \xrightarrow{Ni} CH_3-CH_2-CH_2-CH_2-OH \xrightarrow{H^+,\ heat}$

Butanal

$CH_3-CH_2-CH{=}CH_2 + H_2O \longrightarrow CH_3-CH_2-CH(OH)-CH_3 \xrightarrow{[O]}$

$CH_3-CH_2-C(=O)-CH_3$

Butanone

15.65 **a.** acetal; propanal and methanol
b. hemiacetal; butanone and ethanol
c. acetal; cyclohexanone and ethanol

16

Carbohydrates

Study Goals

- Identify the common carbohydrates in the diet.
- Distinguish between monosaccharides, disaccharides, and polysaccharides.
- Identify the chiral carbons in a carbohydrate.
- Label the Fischer projection for a monosaccharide as the D- or L- enantiomer.
- Write Haworth structures for monosaccharides.
- Describe the structural units and bonds in disaccharides and polysaccharides.

Think About It

1. What are some foods you eat that contain carbohydrates?

2. What elements are found in carbohydrates?

3. What carbohydrates are present in table sugar, milk, and wood?

4. What is meant by a "high-fiber" diet?

Key Terms

Match the following key terms with the descriptions shown below.

A. carbohydrate **B.** glucose **C.** disaccharide **D.** Haworth structure **E.** cellulose

1. ____ A simple or complex sugar composed of a carbon chain with an aldehyde or ketone group and several hydroxyl groups

2. ____ A cyclic structure that represents the closed chain form of a monosaccharide

3. ____ An unbranched polysaccharide that cannot be digested by humans

4. ____ A hexoaldose that is the most prevalent monosaccharide in the diet

5. ____ A carbohydrate that contains two monosaccharides linked by a glycosidic bond

Answers **1.** A **2.** D **3.** E **4.** B **5.** C

16.1 Types of Carbohydrates

- Carbohydrates are classified as monosaccharides (simple sugars), disaccharides (two monosaccharide units), and polysaccharides (many monosaccharide units).

◆ Learning Exercise 16.1A

Complete and balance the equations for the photosynthesis of

1. glucose: ________ + ________ $\longrightarrow C_6H_{12}O_6$ + ________

2. ribose: ________ + ________ $\longrightarrow C_5H_{10}O_5$ + ________

Answers

1. $6CO_2 + 6H_2O \longrightarrow C_6H_{12}O_6 + 6O_2$

2. $5CO_2 + 5H_2O \longrightarrow C_5H_{10}O_5 + 5O_2$

◆ Learning Exercise 16.1B

Indicate the number of monosaccharide units (1, 2, or many) in each of the following carbohydrates.

1. Sucrose, a disaccharide ________ **2.** Cellulose, a polysaccharide ________

3. Glucose, a monosaccharide ________ **4.** Amylose, a polysaccharide ________

5. Maltose, a disaccharide ________

Answers **1.** two **2.** many **3.** one **4.** many **5.** two

16.2 Classification of Monosaccharides

- In a chiral molecule, there is one or more carbon atoms attached to four different atoms or groups.
- Monosaccharides are polyhydroxy aldedhyes (aldoses) or ketones (ketoses).
- Monosaccharides are classified by the number of carbon atoms as *trioses, tetroses, pentoses, or hexoses.*

◆ Learning Exercise 16.2

Identify the following monosaccharides as aldo- or ketotrioses, tetroses, pentoses, or hexoses.

```
1.  CH2OH        2.    H   O       3.     CH2OH      4.   H   O       5.   H   O
     |                  \ //               |               \ //             \ //
    C=O                  C                C=O               C                C
     |                   |                 |                |                |
    CH2OH            H—C—OH          HO—C—H           H—C—OH           H—C—OH
                         |                 |                |                |
                     H—C—OH          HO—C—H          HO—C—H           H—C—OH
                         |                 |                |                |
                    HO—C—H            H—C—OH           H—C—OH           CH2OH
                         |                 |                |
                       CH2OH             CH2OH          H—C—OH
                                                            |
                                                          CH2OH
```

1. ____________________ **2.** ____________________

3. ____________________ **4.** ____________________

5. ____________________

Answers **1.** ketotriose **2.** aldopentose **3.** ketohexose
4. aldohexose **5.** aldotetrose

16.3 D and L Notations from Fischer Projections

- In a chiral molecule, there are four different atoms or groups bonded to a carbon atom.
- The mirror images of a chiral molecule represent two different molecules called enantiomers.
- In a Fischer projection (straight chain), the prefixes D- and L- are used to distinguish between the mirror images. In D-glyceraldehyde, the —OH is on the right of the chiral carbon; it is on the left in L-glyceraldehyde.
- In the Fischer projection of a monosaccharide, the chiral —OH farthest from the carbonyl group (C=O) is on the left side in the L-isomer, and on the right side in the D-isomer.

◆ Learning Exercise 16.3A

State whether each of the following molecules is chiral or not chiral.

1. Cl | H—C—Cl | CH_3 ________
2. Cl | H—C—OH | CH_3 ________
3. CHO | H—C—OH | CH_3 ________

Answers **1.** not chiral **2.** chiral **3.** chiral

◆ Learning Exercise 16.3B

Determine whether each of the following molecules is a chiral compound or not.

1. the central carbon atom is attached to two identical groups ________
2. contains a carbon attached to four different groups ________
3. has identical mirror images ________

Answers **1.** not chiral **2.** chiral **3.** not chiral

◆ Learning Exercise 16.3C

Identify each of the following sugars as the D- or L- isomer.

1. CH_2OH | C=O | HOCH | HCOH | CH_2OH ____-Xylulose
2. CHO | HCOH | HCOH | HOCH | HOCH | CH_2OH ____-Mannos
3. CHO | HOCH | HCOH | CH_2OH ____-Threose
4. CH_2OH | C=O | HOCH | HOCH | CH_2OH ____-Ribulose

Answers **1.** D-Xylulose **2.** L-Mannose **3.** D-Threose **4.** L-Ribulose

◆ Learning Exercise 16.3D

Write the mirror image of each of the sugars in learning exercise 16.3C and give the D- or L-name.

1. **2.** **3.** **4.**

Answers

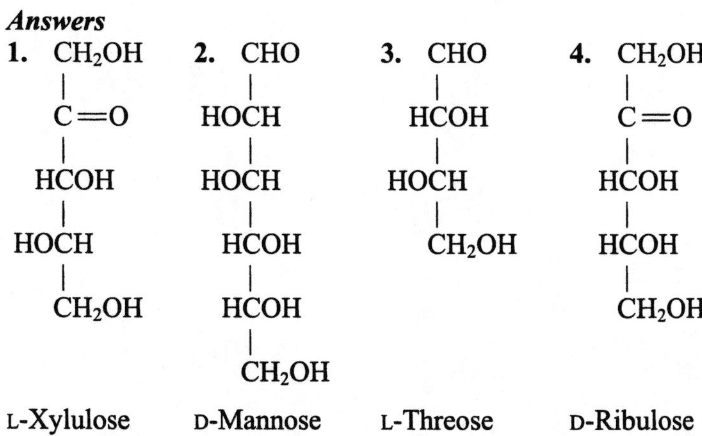

L-Xylulose D-Mannose L-Threose D-Ribulose

16.4 Structures of Some Important Monosaccharides

- Important monosaccharides are the aldohexoses glucose and galactose and the ketohexose fructose.

◆ Learning Exercise 16.4A

Identify the monosaccharide (glucose, fructose, or galactose) that fits the following description.

1. obtained as a hydrolysis product of amylose ____________________
2. also known as fruit sugar ____________________
3. accumulates in the disease known as *galactosemia* ____________________
4. the monosaccharide unit in cellulose ____________________
5. also known as blood sugar ____________________

Answers **1.** glucose **2.** fructose **3.** galactose **4.** glucose **5.** glucose

◆ Learning Exercise 16.4B

Draw the open-chain structure for the following monosaccharides:

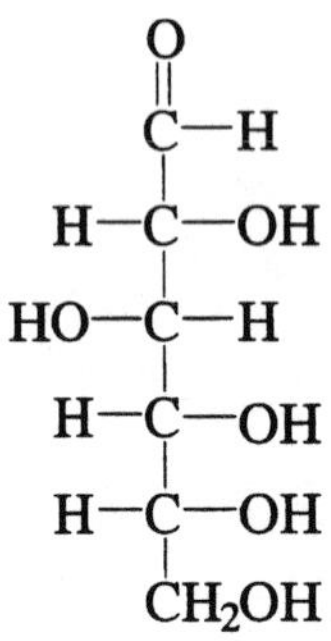

D-Glucose D-Galactose D-Fructose

Answers

```
      O
      ‖
      C—H
      |
   H—C—OH
      |
  HO—C—H
      |
   H—C—OH
      |
   H—C—OH
      |
      CH2OH
   D-glucose
```

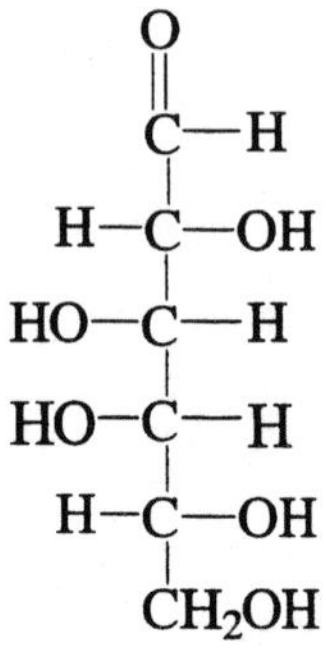

```
      O
      ‖
      C—H
      |
   H—C—OH
      |
  HO—C—H
      |
  HO—C—H
      |
   H—C—OH
      |
      CH2OH
  D-Galactose
```

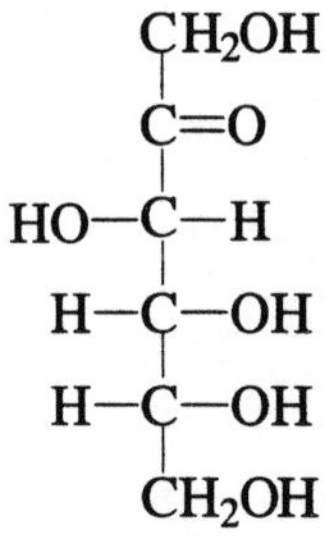

```
      CH2OH
      |
      C=O
      |
  HO—C—H
      |
   H—C—OH
      |
   H—C—OH
      |
      CH2OH
  D-Fructose
```

16.5 Cyclic Structures of Monosaccharides

- The predominant form of monosaccharides is the cyclic form of five or six atoms called the Haworth structure. The cyclic structure forms by a reaction between an OH on carbon 5 in hexoses with the carbonyl group of the same molecule to form a hemiacetal.
- The formation of a new hydroxyl group on carbon 1 (or 2 in fructose) gives α and β anomers of the cyclic monosaccharide. Because the molecule opens and closes continuously (mutarotation) while in solution, both anomers are present.

◆ Learning Exercise 16.5A

Describe each of the following characteristics for D-glucose.

1. The functional groups present in the open chain
2. The functional groups present in the cyclic structure of D-glucose
3. The reason why glucose is a reducing sugar

Answers

1. one aldehyde group, several hydroxyl (OH) groups
2. one hemiacetal group (ether link and alcohol on same carbon) and several hydroxyl groups
3. For the short time that the open chain is present during mutarotation, an aldehyde group becomes available for oxidation.

◆ Learning Exercise 16.5B

Write the Haworth structures (α-anomers) for the following.

1. D-Glucose **2.** D-Galactose **3.** D-Fructose

Answers

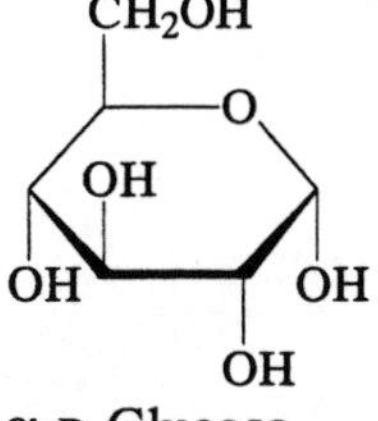

α-D-Glucose

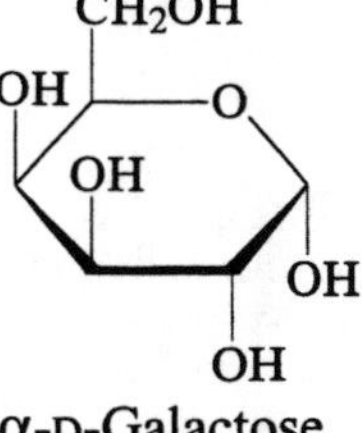

α-D-Galactose

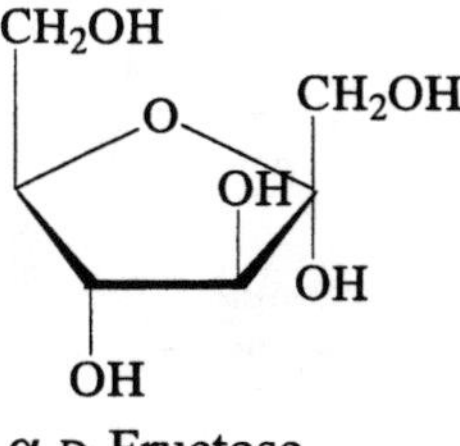

α-D-Fructose

16.6 Chemical Properties of Monosaccharides

- Monosaccharides contain functional groups that undergo oxidation or reduction.
- Monosaccharides are called *reducing sugars* because the aldehyde group (also available in ketoses) is oxidized by a metal ion such as Cu^{2+} in Benedict's solution, which is reduced.
- Monosaccharides are also reduced to give sugar alcohols.

◆ Learning Exercise 16.6

What changes occur when a reducing sugar reacts with Benedict's reagent?

Answers The carbonyl group of the reducing sugar is oxidized to a carboxylic acid group; the Cu^{2+} ion in Benedict's reagent is reduced to Cu^{+}, which forms a brick-red solid of Cu_2O.

16.7 Disaccharides

- Disaccharides are glycosides, two monosaccharide units joined together by a glycosidic bond:

$$\text{monosaccharide(1)} + \text{monosaccharide (2)} \longrightarrow \text{disaccharide} + H_2O$$

- In the most common disaccharides, maltose, lactose, and sucrose, there is at least one glucose unit.
- In the disaccharide maltose, two glucose units are linked by an α-1,4-glycosidic bond. The α-1,4 indicates that the —OH of the alpha anomer at carbon 1 was bonded to the —OH on carbon 4 of the other glucose molecule.
- When a disaccharide is hydrolyzed by water, the products are a glucose unit and one other monosaccharide.

$$\text{Maltose} + H_2O \longrightarrow \text{Glucose} + \text{Glucose}$$
$$\text{Lactose} + H_2O \longrightarrow \text{Glucose} + \text{Galactose}$$
$$\text{Sucrose} + H_2O \longrightarrow \text{Glucose} + \text{Fructose}$$

◆ Learning Exercise 16.7

a. What is a glycosidic bond?

b. For the following disaccharides, state (a) the monosaccharide units, (b) the type of glycosidic bond, and (c) the name of the disaccharide.

1. CH_2OH CH_2OH O O OH OH OH O HO OH OH

2. CH_2OH CH_2OH O HO O OH O OH OH OH OH

	a. Monosaccharide(s)	**b.** Type of glycosidic bond	**c.** Name of dissacharide
1.			
2.			

3.

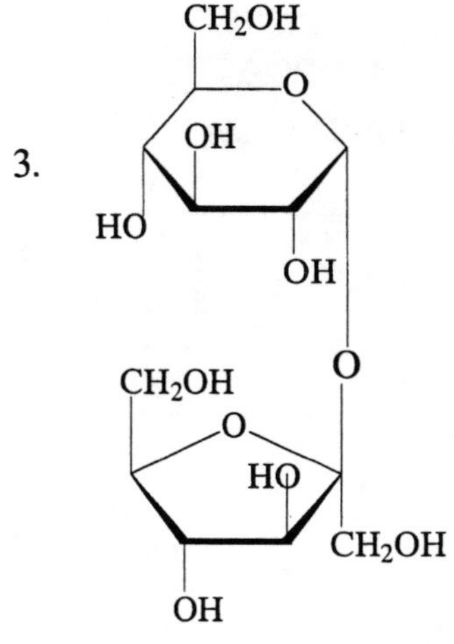

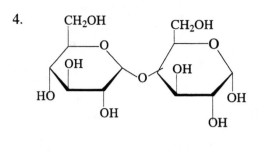

4.

	a. Monosaccharide units	**b.** Type of glycosidic bond	**c.** Name of disaccharide
3.			
4.			

Answers

a. A glycosidic bond forms between the OH of the hemiacetal group of a sugar with the —OH of another compound, usually another sugar.

b.
1. **(a)** two glucose units **(b)** α-1,4-glycosidic bond **(c)** β-maltose
2. **(a)** galactose + glucose **(b)** β-1,4-glycosidic bond **(c)** α-lactose
3. **(a)** fructose + glucose **(b)** α-1, β-2 glycosidic bond **(c)** sucrose
4. **(a)** two glucose units **(b)** α-1,4-glycosidic bond **(c)** α-maltose

16.8 Polysaccharides

- Polysaccharides are polymers of monosaccharide units.
- Starches consist of amylose, an unbranched chain of glucose; amylopectin is a branched polymer of glucose. Glycogen, the storage form of glucose in animals, is similar to amylopectin with more branching.
- Cellulose is also a polymer of glucose, but in cellulose the glycosidic bonds are β bonds rather than α bonds as in the starches. Humans can digest starches to obtain energy, but not cellulose. However, cellulose is important as a source of fiber in our diets.

◆ Learning Exercise 16.8

List the monosaccharides and describe the glycosidic bonds in each of the following carbohydrates:

	Monosaccharides	Type(s) of glycosidic bonds
1. amylose	________________	________________
2. amylopectin	________________	________________
3. glycogen	________________	________________
4. cellulose	________________	________________

Answers **1.** glucose; α-1,4-glycosidic bonds **2.** glucose; α-1,4- and α-1,6-glycosidic bonds
3. glucose; α-1,4- and α-1,6-glycosidic bonds **4.** glucose; β-1,4-glycosidic bonds

Check List for Chapter 16

You are ready to take the practice test for Chapter 16. Be sure that you have accomplished the following learning goals for this chapter. If you are not sure, review the section listed at the end of the goal. Then apply your new skills and understanding to the practice test.

After studying Chapter 16, I can successfully:

____ Classify carbohydrates as monosaccharides, disaccharides, and polysaccharides (16.1).

____ Classify a monosaccharide as an aldose or ketose and indicate the number of carbon atoms (16.2).

____ Draw and identify D- and L- Fischer projections for carbohydrate molecules (16.3).

____ Draw the open-chain structures for D-glucose, D-galactose, and D-fructose (16.4).

____ Draw or identify the cyclic structures of monosaccharides (16.5).

____ Describe some chemical properties of carbohydrates (16.6).

____ Describe the monosaccharide units and linkages in disaccharides (16.7).

____ Describe the structural features of amylose, amylopectin, glycogen, and cellulose (16.8).

Practice Test for Chapter 16

1. The requirements for photosynthesis are

A. sun. **B.** sun and water. **C.** water and carbon dioxide.
D. sun, water, and carbon dioxide. **E.** carbon dioxide and sun.

2. What are the products of photosynthesis?

A. carbohydrates **B.** carbohydrates and oxygen
C. carbon dioxide and oxygen **D.** carbohydrates and carbon dioxide
E. water and oxygen

3. The name "carbohydrate" came from the fact that

A. carbohydrates are hydrates of water.
B. carbohydrates contain hydrogen and oxygen in a 2:1 ratio.
C. carbohydrates contain a great quantity of water.
D. all plants produce carbohydrates.
E. carbon and hydrogen atoms are abundant in carbohydrates.

4. What functional group are in the open chains of monosaccharides?

A. hydroxyl groups
B. aldehyde groups
C. ketone groups
D. hydroxyl and aldehyde or ketone groups
E. hydroxyl groups and a hemiacetal group

5. What is the classification of the following sugar?

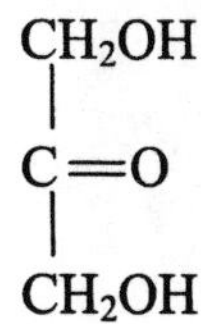

A. aldotriose **B.** ketotriose **C.** aldotetrose **D.** ketotetrose **E.** ketopentose

Questions 6 through 10 refer to

CH_2OH
O OH
OH
HO
OH

6. It is the cyclic structure of an

A. aldotriose. **B.** ketopentose. **C.** aldopentose.
D. aldohexose. **E.** aldoheptose.

7. This is a Haworth structure of

A. fructose. **B.** glucose. **C.** ribose.
D. glyceraldehyde. **E.** galactose.

8. It is at least one of the products of the complete hydrolysis of

A. maltose. **B.** sucrose. **C.** lactose.
D. glycogen. **E.** all of these

9. A Benedict's test with this sugar would

A. be positive. **B.** be negative. **C.** produce a blue precipitate.
D. give no color change. **E.** produce a silver mirror.

10. It is the monosaccharide unit used to build polymers of

A. amylose. **B.** amylopectin. **C.** cellulose.
D. glycogen **E.** all of these

Identify each of the carbohydrates described in 11 through 15 as:

A. maltose **B.** sucrose **C.** cellulose
D. amylopectin **E.** glycogen

11. _____ A disaccharide that is not a reducing sugar

12. _____ A disaccharide that occurs as a breakdown product of amylose

13. _____ A carbohydrate that is produced as a storage form of energy in plants

14. _____ The storage form of energy in humans

15. _____ A carbohydrate that is used for structural purposes by plants

For questions 16 through 20 select answers from:

A. amylose **B.** cellulose **C.** glycogen **D.** lactose **E.** sucrose

16. _____ A polysaccharide composed of many glucose units linked by α-1,4-glycosidic bonds.

17. _____ A sugar containing both glucose and galactose.

18. _____ A sugar composed of glucose units joined by both α-1,4- and α-1,6-glycosidic bonds.

19. _____ A sugar that has no anomeric forms.

20. _____ A carbohydrate composed of glucose units joined by β-1,4-glycosidic bonds.

For questions 21 through 25, select ***answers*** from the following:

A. amylose **B.** lactose **C.** sucrose **D.** maltose

21. _____ A sugar composed of glucose and fructose

22. _____ Gives a positive Benedict's test, but is negative for fermentation

23. _____ Gives a negative Benedict's test and a positive fermentation test

24. _____ Gives a positive iodine test

25. _____ Gives galactose upon hydrolysis

Answers to the Practice Test

1. D	**2.** B	**3.** B	**4.** D	**5.** B
6. D	**7.** B	**8.** E	**9.** A	**10.** E
11. B	**12.** A	**13.** D	**14.** E	**15.** C
16. A	**17.** D	**18.** C	**19.** E	**20.** B
21. C	**22.** B	**23.** C	**24.** A	**25.** B

Answers and Solutions to Selected Text Problems

16.1 Photosynthesis requires CO_2, H_2O, and the energy from the sun. Respiration requires O_2 from the air and glucose from our foods.

16.3 A monosaccharide is a simple sugar composed of three to six carbon atoms. A disaccharide is composed of two monosaccharide units.

16.5 Hydroxyl groups and a carbonyl are found in all monosaccharides.

16.7 The name ketopentose tells us that the compound contains a ketone functional group and has five carbon atoms. In addition, all monosaccharides contain hydroxyl groups.

16.9
a. This monosaccharide is a ketose; it has a carbonyl between two carbon atoms.
b. This monosaccharide is an aldose; it has a CHO, an aldehyde group.
c. This monosaccharide is a ketose; it has a carbonyl between two carbon atoms.
d. This monosaccharide is an aldose; it has a CHO, an aldehyde group.
e. This monosaccharide is an aldose; it has a CHO, an aldehyde group.

16.11 A Fischer projection is a two-dimensional representation of the three-dimensional structure of a molecule.

16.13
a. This structure is a D-isomer since the hydroxyl on the chiral carbon farthest from the carbonyl is on the right.
b. This structure is a D-isomer since the hydroxyl on the chiral carbon farthest from the carbonyl is on the right.
c. This structure is an L-isomer since the hydroxyl on the chiral carbon farthest from the carbonyl is on the left.
d. This structure is a D-isomer since the hydroxyl on the chiral carbon farthest from the carbonyl is on the right.

16.15

a.
```
      CHO
  H ──┼── OH
 HO ──┼── H
      CH2OH
```

b.
```
      CH2OH
      ══ O
  H ──┼── OH
 HO ──┼── H
      CH2OH
```

c.
```
      CHO
 HO ──┼── H
 HO ──┼── H
  H ──┼── OH
  H ──┼── OH
      CH2OH
```

d.
```
      CHO
 HO ──┼── H
 HO ──┼── H
 HO ──┼── H
 HO ──┼── H
      CH2OH
```

16.17 L-glucose is the mirror image of D-glucose.

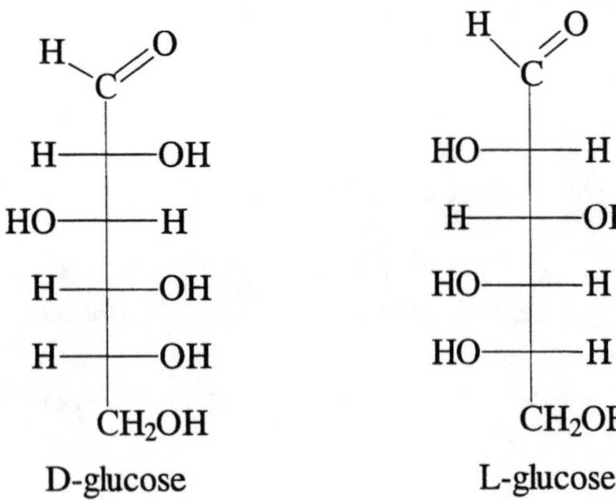

D-glucose

H O
C
HO—H
H—OH
HO—H
HO—H
CH_2OH

L-glucose

16.19 In D-galactose the hydroxyl on carbon four extends to the left; in glucose this hydroxyl group goes to the right.

16.21
a. Glucose is also called blood sugar.
b. Galactose is not metabolized in the condition called galactosemia.
c. Another name for fructose is fruit sugar.

16.23 In the cyclic structure of glucose, there are five carbon atoms and an oxygen atom in the ring.

16.25 In the α anomer, the hydroxyl (—OH) on carbon 1 is down; the β anomer has the hydroxyl (—OH) on carbon 1 up;

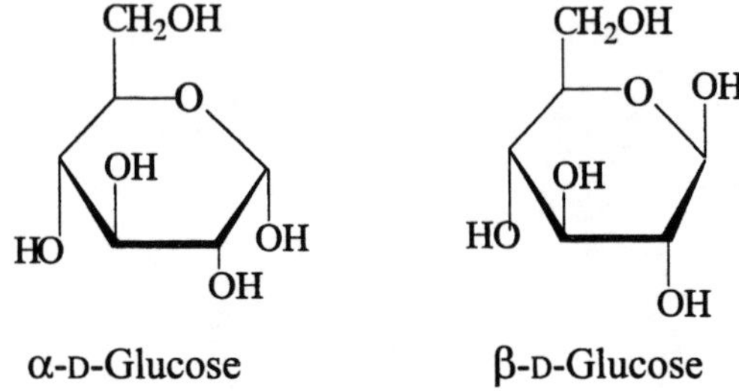

α-D-Glucose

CH_2OH
O OH
OH
HO
OH

β-D-Glucose

16.27
a. This is the α-anomer because the —OH on carbon 1 is down.
b. This is the α-anomer because the —OH on carbon 1 is down.

16.29

CH_2OH
H—C—OH
HO—C—H
H—C—OH
CH_2OH

Xylitol

16.31 Oxidation product:

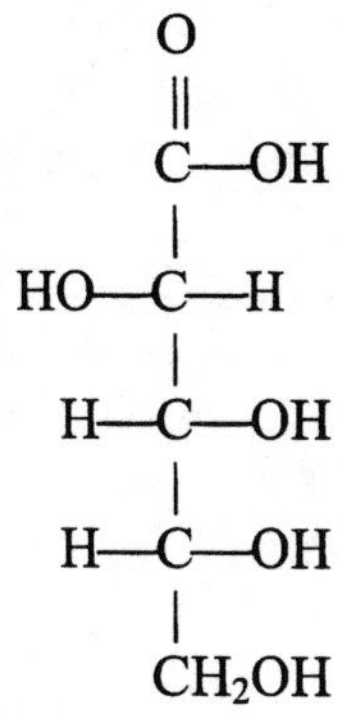

```
      O
      ‖
      C—OH
      |
 HO—C—H
      |
  H—C—OH
      |
  H—C—OH
      |
      CH2OH
```

Reduction product:

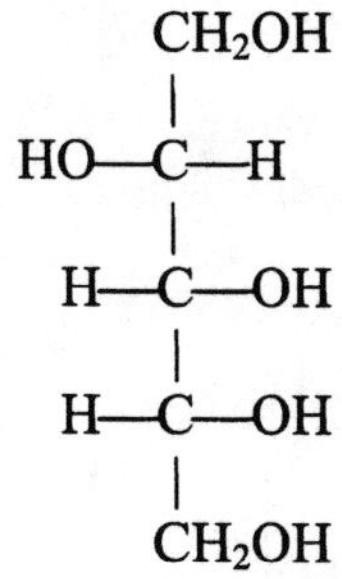

```
      CH2OH
      |
 HO—C—H
      |
  H—C—OH
      |
  H—C—OH
      |
      CH2OH
```

D-arabitol

16.33

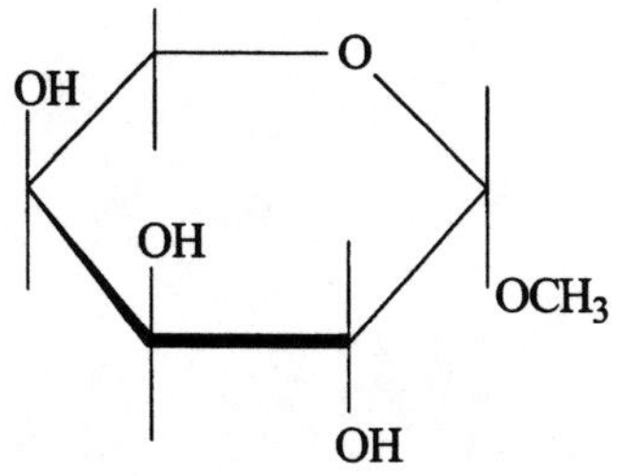

OH O OH OCH$_3$ OH

α-anomer

OH O OCH$_3$ OH OH

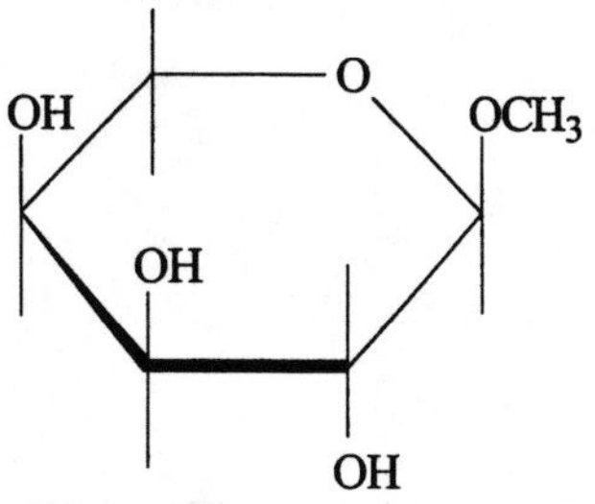

β-anomer

16.35 **a.** When this disaccharide is hydrolyzed, galactose and glucose are produced. The glycosidic bond is a β-1,4 bond since the ether bond is up from the 1 carbon of the galactose, which is on the left in the drawing to the 4 carbon of the glucose on the right. β-lactose is the name of this disaccharide since the free hydroxyl is up.

b. When this disaccharide is hydrolyzed, two molecules of glucose are produced. The glycosidic bond is an α-1,4 bond since the ether bond is down from the 1 carbon of the glucose on the left to the 4 carbon of the glucose on the right. α-maltose is the name of this disaccharide since the free hydroxyl is down.

16.37 **a.** Will undergo mutarotation; can be oxidized
b. Will undergo mutarotation; can be oxidized

16.39 **a.** Another name for table sugar is sucrose.
b. Lactose is the disaccharide found in milk and milk products.
c. Maltose is also called malt sugar.
d. When lactose is hydrolyzed, the products are the monosaccahrides galactose and glucose.

16.41 **a.** Amylose is an unbranched polymer of glucose units joined by α-1,4 bonds; amylopectin is a branched polymer of glucose joined by α-1,4 and α-1,6 bonds.

b. Amylopectin, produced by plants, is a branched polymer of glucose joined by α-1,4 and α-1,6 bonds. Glycogen, which is made by animals, is a highly branched polymer of glucose joined by α-1,4 and α-1,6 bonds.

16.43 **a.** Cellulose is not digestible by humans since we do not have the enzymes necessary to break the β -1,4-glycosidic bonds in cellulose.

b. Amylose and amylopectin are the storage forms of carbohydrates in plants.

c. Amylose is the polysaccharide, which contains only α-1,4 glycosidic bonds.

d. Glycogen contains many α-1,4 and α-1,6 bonds and is the most highly branched polysaccharide.

16.45 They differ only at carbon 4; the —OH in D-glucose is on the right side and in D-galactose it is on the left side.

16.47 D-galactose is the mirror image of L-galactose. In D-galactose, the —OH group on carbon 5 is on the right side whereas in L-galactose, the —OH group on carbon 5 is on the left side.

16.49 **a.**

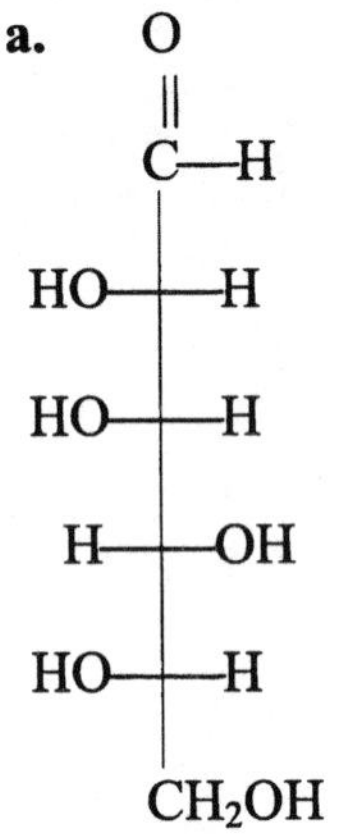

O
||
C—H
HO—H
HO—H
H—OH
HO—H
CH_2OH

L-Gulose

b.

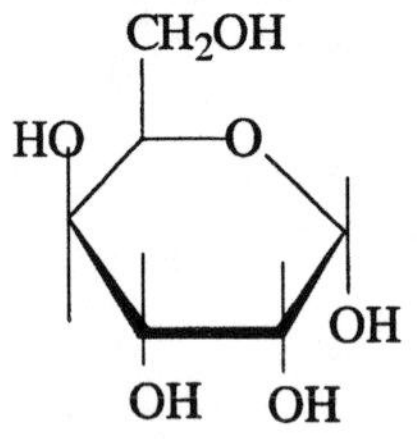

CH_2OH, HO, O, OH, OH, OH

α-D-Gulose

CH_2OH, HO, O, OH, OH, OH

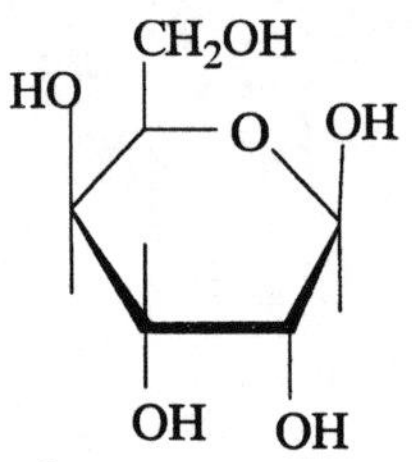

β-D-Gulose

16.51 Since sorbitol can be oxidized to D-glucose, it must contain the same number of carbons with the same groups attached as glucose. The difference is that sorbitol has only hydroxyls while glucose has an aldehyde group. In sorbitol, the aldehyde group is changed to a hydroxyl.

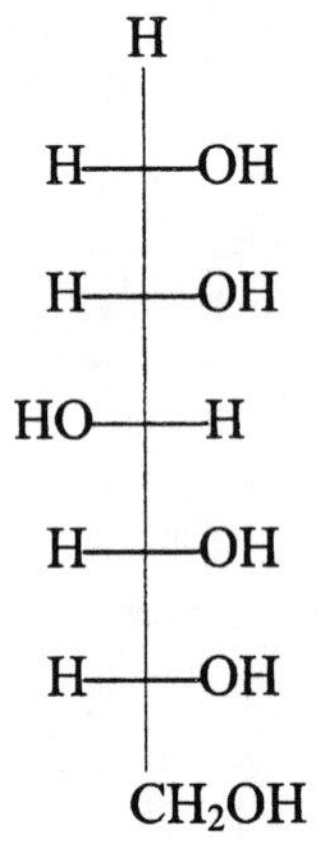

H
H—OH
H—OH
HO—H
H—OH
H—OH
CH_2OH

This hydroxyl is an aldehyde in glucose.

16.53 The α-galactose forms an open chain structure and when the chain closes, it can form both α- and β-galactose.

16.55

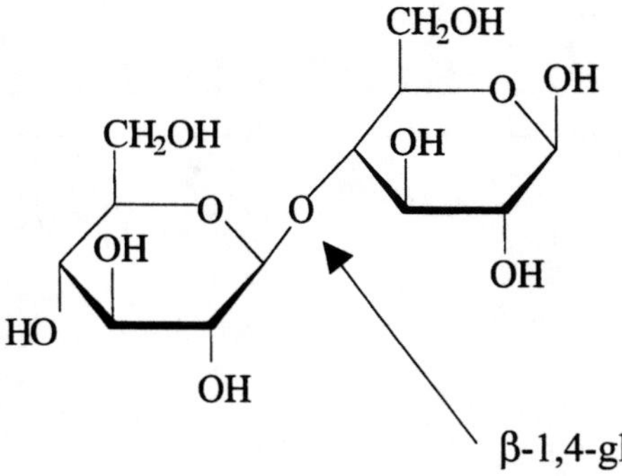

β-1,4-glycosidic bond. The bond from the glucose on the left is up (β).

16.57 **a.**

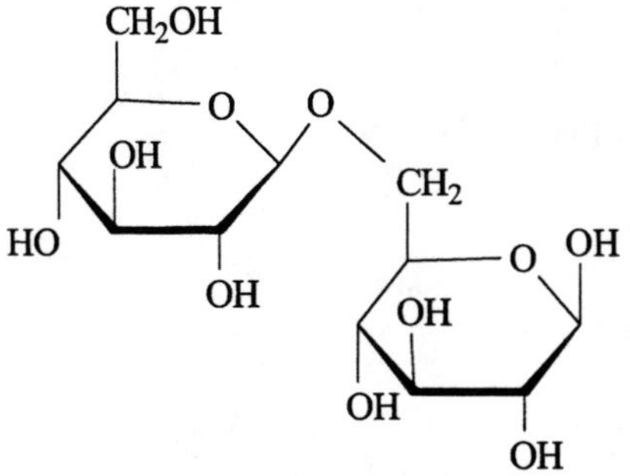

b. Yes. The hemiacetal on the right side can open up to form the open chain with an aldehyde.

17

Carboxylic Acids and Esters

Study Goals

- Name and write structural formulas of carboxylic acids and esters.
- Describe the boiling points and solubility of carboxylic acids.
- Write equations for the ionization of carboxylic acids in water.
- Write equations for the esterification, hydrolysis, and saponification of esters.

Think About It

1. Why do vinegar and citrus juices taste sour?

2. What type of compound gives flowers and fruits their pleasant aromas?

Key Terms

Match the key term with the correct statement shown below.

a. carboxylic acid **b.** saponification **c.** esterification
d. hydrolysis **e.** ester

1. _____ An organic compound containing the carboxyl group (—COOH)

2. _____ A reaction of a carboxylic acid and an alcohol in the presence of an acid catalyst

3. _____ A type of organic compound that produces pleasant aromas in flowers and fruits

4. _____ The hydrolysis of an ester with a strong base producing a salt of the carboxylic acid and an alcohol

5. _____ The splitting of a molecule such as an ester by the addition of water in the presence of an acid

Answers **1.** a **2.** c **3.** e **4.** b **5.** d

17.1 Carboxylic Acids

- In the IUPAC system, a carboxylic acid is named by replacing the *ane* ending with *oic acid.* Simple acids usually are named by the common naming system using the prefixes: **form** (1C), **acet** (2C), **propion** (4C), **butyr** (4C), followed by ***ic acid.***

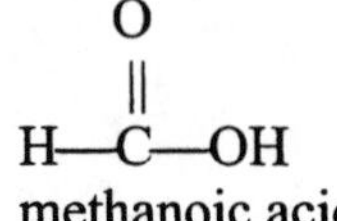
methanoic acid
(formic acid)

$CH_3—\overset{\overset{O}{\|}}{C}—OH$
ethanoic acid
(acetic acid)

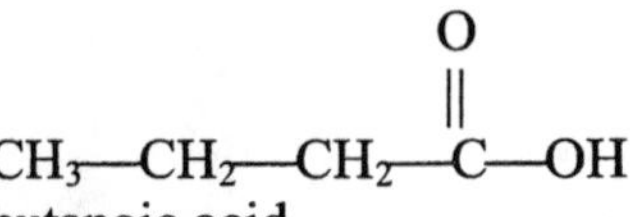
butanoic acid
(butyric acid)

- A carboxylic acid is prepared by the oxidation of a primary alcohol or an aldehyde.

$$CH_3—CH_2—OH \xrightarrow{[O]} CH_3—\overset{\overset{O}{\|}}{C}—H \xrightarrow{[O]} CH_3—\overset{\overset{O}{\|}}{C}—OH$$

◆ Learning Exercise 17.1A

Give the IUPAC and common names for each of the following carboxylic acids.

1. 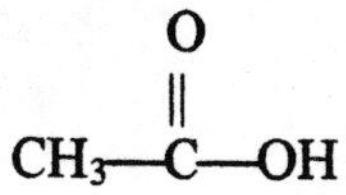 $CH_3—\overset{\overset{\large O}{\|}}{C}—OH$

2. $CH_3—\overset{\overset{\large OH}{|}}{CH}—\overset{\overset{\large O}{\|}}{C}—OH$

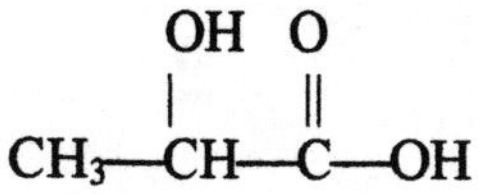

3. $CH_3—\overset{\overset{\large CH_3}{|}}{CH}—CH_2—\overset{\overset{\large O}{\|}}{C}—OH$

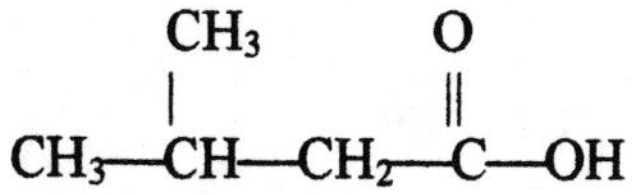

4. 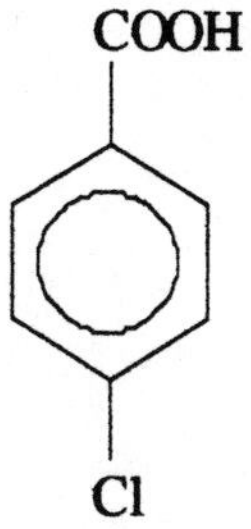 (benzene ring with COOH at top and Cl at bottom, para positions)

Answers

1. ethanoic acid (acetic acid)
2. 2-hydroxypropanoic acid (α-hydroxypropionic acid)
3. 3-methylbutanoic acid (β-methylbutyric acid)
4. 4-chlorobenzoic acid (*p*-chlorobenzoic acid)

◆ Learning Exercise 17.1B

A. Write the formulas for the following carboxylic acids.

1. acetic acid
2. 2-ketobutanoic acid
3. benzoic acid
4. β-hydroxypropionic acid
5. formic acid
6. 3-methylpentanoic acid

Answers

1. $CH_3-\overset{\overset{O}{\|}}{C}-OH$

2. $CH_3-CH_2-\overset{\overset{O}{\|}}{C}-\overset{\overset{O}{\|}}{C}-OH$

3. $C_6H_5-\overset{\overset{O}{\|}}{C}-OH$ (benzene ring)

4. $HO-CH_2-CH_2-\overset{\overset{O}{\|}}{C}-OH$

5. $H-\overset{\overset{O}{\|}}{C}-OH$

6. $CH_3-CH_2-\overset{\overset{CH_3}{|}}{CH}-CH_2-\overset{\overset{O}{\|}}{C}-OH$

◆ Learning Exercise 17.1C

Write the structural formula of the appropriate aldehyde to produce the following.

1. propanoic acid

2. β-methylbutyric acid

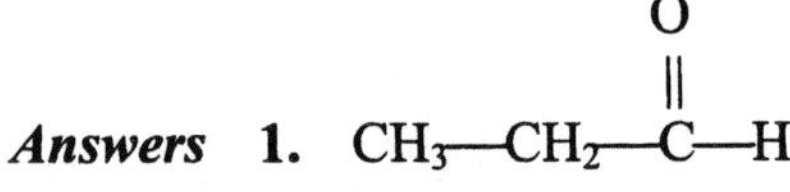

2. $CH_3-\overset{\overset{CH_3}{|}}{CH}-CH_2-\overset{\overset{O}{\|}}{C}-H$

17.2 Physical Properties of Carboxylic Acids

- Carboxylic acids have higher boiling points than other polar compounds such as alcohols.
- Because they have two polar groups, two carboxylic acids form a dimer, which contains two sets of hydrogen bonds.
- Carboxylic acids with one to four carbon atoms are very soluble in water.

◆ Learning Exercise 17.2A

Indicate whether each of the following carboxylic acids are soluble in water.

1. ____ hexanoic acid **2.** ____ acetic acid **3.** ____ propanoic acid

4. ____ benzoic acid **5.** ____ formic acid **6.** ____ octanoic acid

Answers **1.** no **2.** yes **3.** yes **4.** no **5.** yes **6.** no

◆ Learning Exercise 17.2B

Identify the compound in each pair that has the higher boiling point.

1. acetic acid or butyric acid
2. propanoic acid or 2-propanol
3. propanoic acid or propanone
4. acetic acid or acetaldehyde

Answers
1. butyric acid
2. propanoic acid
3. propanoic acid
4. acetic acid

17.3 Acidity of Carboxylic Acids

- As weak acids, carboxylic acids ionize slightly in water to form acidic solutions of H_3O^+ and a carboxylate ion.
- When bases neutralize carboxylic acids, the products are carboxylic acid salts and water.

◆ Learning Exercise 17.3A

Write the products for the ionization of the following carboxylic acids in water.

1. $CH_3—CH_2—\overset{\overset{\large O}{\|}}{C}—OH + H_2O \rightleftarrows$

2. benzoic acid + $H_2O \rightleftarrows$

Answers

1. $CH_3—CH_2—\overset{\overset{\large O}{\|}}{C}—O^- + H_3O^+$

2. 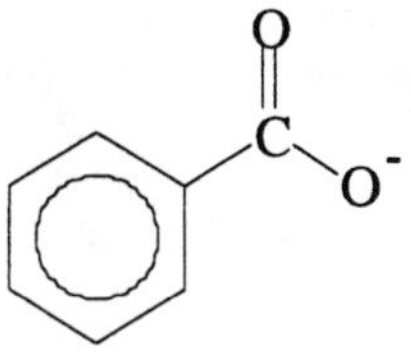 $C_6H_5—\overset{\overset{\large O}{\|}}{C}—O^-$ + H_3O^+

◆ Learning Exercise 17.3B

Identify the carboxylic acid that produces more H_3O^+ in water.

1. formic acid (K_a 1.8×10^{-4}) and acetic acid ($K_a = 1.8 \times 10^{-5}$)

2. acetic acid ($K_a = 1.8 \times 10^{-5}$) and benzoic acid ($K_a = 6.5 \times 10^{-5}$)

Answers
1. formic acid (larger K_a)
2. benzoic acid (larger K_a)

◆ Learning Exercise 17.3C

Write the products and names for each of the following reactions.

1. $CH_3{-}CH_2{-}\overset{\overset{\large O}{\|}}{C}{-}OH + NaOH \longrightarrow$

2. formic acid + KOH ⟶

Answers

1. $CH_3{-}CH_2{-}\overset{\overset{\large O}{\|}}{C}{-}O^-Na^+ + H_2O$
Sodium propanoate
(sodium propionate)

2. $H{-}\overset{\overset{\large O}{\|}}{C}{-}O^-K^+ + H_2O$
potassium methanoate
(potassium formate)

17.4 Esters of Carboxylic Acids

- In the presence of a strong acid, carboxylic acids react with alcohols to produce esters and water.

◆ Learning Exercise 17.4A

Write the products of the following reactions.

1. $CH_3{-}\overset{\overset{\large O}{\|}}{C}{-}OH + CH_3{-}OH \xrightarrow{H^+}$

2. $H{-}\overset{\overset{\large O}{\|}}{C}{-}OH + CH_3{-}CH_2{-}OH \xrightarrow{H^+}$

3. $C_6H_5{-}\overset{\overset{\large O}{\|}}{C}{-}OH + HO{-}CH_3 \xrightarrow{H^+}$

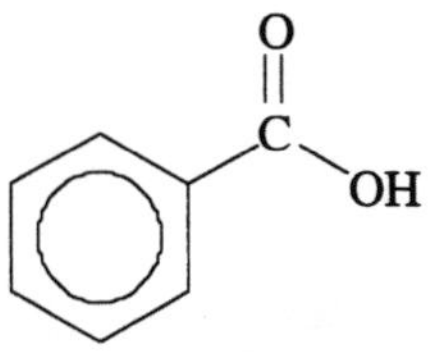

4. propanoic acid and ethanol $\xrightarrow{H^+}$

Answers

1. $CH_3{-}\overset{\overset{\large O}{\|}}{C}{-}O{-}CH_3 + H_2O$

2. $H{-}\overset{\overset{\large O}{\|}}{C}{-}O{-}CH_2{-}CH_3 + H_2O$

3. $C_6H_5{-}\overset{\overset{\large O}{\|}}{C}{-}O{-}CH_3 + H_2O$

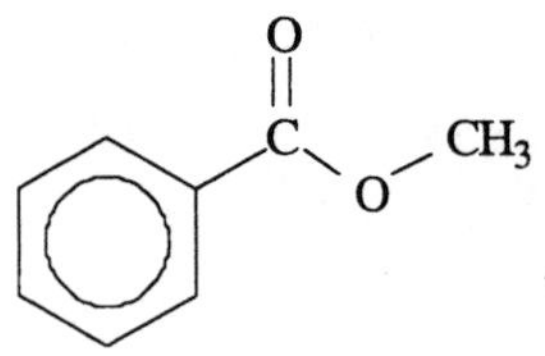

4. $CH_3{-}CH_2{-}\overset{\overset{\large O}{\|}}{C}{-}O{-}CH_2{-}CH_3 + H_2O$

17.5 Naming Esters

- The names of esters consist of two words, one from the alcohol and the other from the carboxylic acid with the *ic* ending replaced by *ate.*

$$CH_3-\overset{\overset{\displaystyle O}{\|}}{C}-O-CH_3$$ methyl ethanoate (IUPAC) or methyl acetate (common)

◆ Learning Exercise 17.5A

Name each of the following esters.

1. $CH_3-\overset{\overset{\displaystyle O}{\|}}{C}-O-CH_2-CH_3$

2. $CH_3-CH_2-CH_2-\overset{\overset{\displaystyle O}{\|}}{C}-O-CH_3$

3. $CH_3-CH_2-\overset{\overset{\displaystyle O}{\|}}{C}-O-CH_2-\overset{\overset{\displaystyle OH}{|}}{CH}-CH_3$

4. 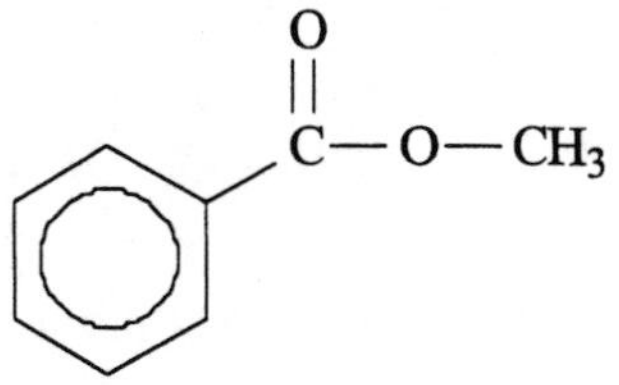 $C_6H_5-\overset{\overset{\displaystyle O}{\|}}{C}-O-CH_3$ (benzene ring)

Answers

1. ethyl ethanoate (ethyl acetate)
2. methylbutanoate (methylbutyrate)
3. 2-hydroxypropylpropanoate (2-hydroxypropylpropionate)
4. methylbenzoate

◆ Learning Exercise 17.5B

Write structural formulas for each of the following esters.

1. propylacetate
2. ethylbutyrate
3. ethylpropanoate
4. ethylbenzoate

Answers **1.** $CH_3-\overset{\overset{\displaystyle O}{\|}}{C}-O-CH_2-CH_2-CH_3$ **2.** $CH_3-CH_2-CH_2-\overset{\overset{\displaystyle O}{\|}}{C}-O-CH_2-CH_3$

3. $CH_3-CH_2-\overset{\overset{\displaystyle O}{\|}}{C}-O-CH_2-CH_3$ **4.** 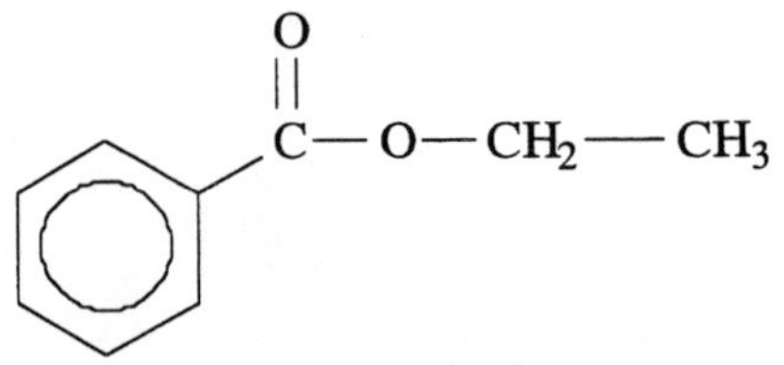 $C_6H_5-\overset{\overset{\displaystyle O}{\|}}{C}-O-CH_2-CH_3$

17.6 Properties of Esters

- Esters shave higher boiling points than alkanes, but lower than alcohols and carboxylic acids of similar mass.
- In hydrolysis, esters are split apart by a reaction with water. When the catalyst is an acid, the products are a carboxylic acid and an alcohol.

$$\underset{\textit{methyl acetate}}{CH_3-\overset{\overset{\displaystyle O}{\|}}{C}-O-CH_3} + H_2O \xrightarrow{H^+} \underset{\textit{acetic acid}}{CH_3-\overset{\overset{\displaystyle O}{\|}}{C}-OH} + \underset{\textit{methyl alcohol}}{HO-CH_3}$$

- Saponification is the hydrolysis of an ester in the presence of a base, which produces a carboxylate salt and an alcohol.

$$\underset{\textit{methyl acetate}}{CH_3-\overset{\overset{\displaystyle O}{\|}}{C}-O-CH_3} + NaOH \longrightarrow \underset{\textit{sodium acetate}}{CH_3-\overset{\overset{\displaystyle O}{\|}}{C}-O^-Na^+} + \underset{\textit{methyl alcohol}}{HO-CH_3}$$

- In saponification, long-chain fatty acids from fats react with strong bases to produce salts of the fatty acids, which are soaps.

◆ Learning Exercise 17.6A

Identify the compound with the higher boiling point in each of the following pairs of compounds.

1. CH_3-CH_2-OH or $H-\overset{\overset{\displaystyle O}{\|}}{C}-O-CH_3$
2. $CH_3-\overset{\overset{\displaystyle O}{\|}}{C}-O-CH_3$ or $CH_3-\overset{\overset{\displaystyle OH}{|}}{CH}-CH_2-CH_3$
3. $CH_3-\overset{\overset{\displaystyle O}{\|}}{C}-O-CH_3$ or $CH_3-CH_2-CH_2-CH_2-CH_3$

Answers **1.** CH_3-CH_2-OH **2.** $CH_3-\overset{\overset{\displaystyle OH}{|}}{CH}-CH_2-CH_3$

3. $CH_3-\overset{\overset{\displaystyle O}{\|}}{C}-O-CH_3$

◆ Learning Exercise 17.6B

Write the products of hydrolysis or saponification for the following esters:

1. $CH_3{-}CH_2{-}CH_2{-}\overset{\overset{\large O}{\|}}{C}{-}O{-}CH_3 + H_2O \xrightarrow{H^+}$

2. $CH_3{-}\overset{\overset{\large O}{\|}}{C}{-}O{-}CH_3 + NaOH \longrightarrow$

3. $C_6H_5{-}\overset{\overset{\large O}{\|}}{C}{-}O{-}CH_2{-}CH_3 + KOH \longrightarrow$

4. $C_6H_5{-}\overset{\overset{\large O}{\|}}{C}{-}O{-}CH_2{-}CH_2{-}CH_3 + H_2O \xrightarrow{H^+}$

Answers

1. $CH_3{-}CH_2{-}CH_2{-}\overset{\overset{\large O}{\|}}{C}{-}OH + HO{-}CH_3$

2. $CH_3{-}\overset{\overset{\large O}{\|}}{C}{-}O^- Na^+ + CH_3{-}OH$

3. $C_6H_5{-}\overset{\overset{\large O}{\|}}{C}{-}O^- K^+ + HO{-}CH_2{-}CH_3$

4. $C_6H_5{-}\overset{\overset{\large O}{\|}}{C}{-}OH + HO{-}CH_2{-}CH_2{-}CH_3$

Check List for Chapter 17

You are ready to take the practice test for Chapter 17. Be sure that you have accomplished the following learning goals for this chapter. If you are not sure, review the section listed at the end of the goal. Then apply your new skills and understanding to the practice test.

After studying Chapter 17, I can successfully:

____ Write the IUPAC and common names and draw condensed structural formulas of carboxylic acids (17.1).

____ Describe the solubility and ionization of carboxylic acids in water (17.2).

____ Describe the behavior of carboxylic acids as weak acids and write the structural formulas for the products of neutralization (17.3).

____ Write equations for the preparation of esters (17.4).

____ Write the IUPAC or common names, and condensed structural formulas of esters (17.5).

____ Write equations for the hydrolysis and saponification of esters (17.6).

Practice Test for Chapter 17

Match each structure to its functional group.

A. alcohol **B.** aldehyde **C.** carboxylic acid **D.** ester **E.** ketone

1. ____ 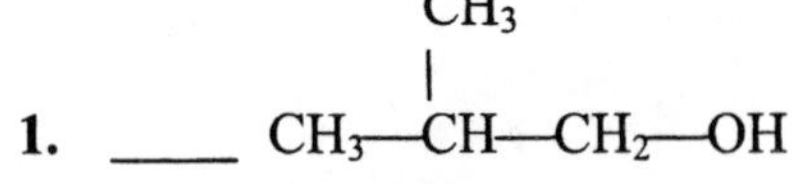
$CH_3-CH(CH_3)-CH_2-OH$

2. ____ $CH_3-CH_2-C(=O)-H$

3. ____ $CH_3-C(=O)-CH_2-CH_3$

4. ____ $CH_3-CH_2-C(=O)-OH$

5. ____ $CH_3-C(=O)-O-CH_3$

Match the names of the following compounds with their structures.

A. $CH_3-C(=O)-O-CH_2-CH_3$ **B.** $CH_3-CH_2-CH_2-C(=O)-O^- Na^+$ **C.** $CH_3-C(=O)-O^- Na^+$

D. $CH_3-CH_2-CH(CH_3)-C(=O)-OH$ **E.** $CH_3-CH_2-C(=O)-O-CH_3$

6. ____ α-methylbutyric acid

7. ____ methylpropanoate

8. ____ sodium butanoate

9. ____ ethyl acetate

10. ____ sodium acetate

11. An aldehyde can be oxidized to give a(n)

A. alcohol. **B.** ketone. **C.** carboxylic acid.
D. ester. **E.** no reaction

12. What is the product when a carboxylic acid reacts with sodium hydroxide?

A. carboxylic acid salt **B.** alcohol **C.** ester
D. aldehyde **E.** no reaction

13. Carboxylic acids are water soluble due to their

A. nonpolar nature. **B.** ionic bonds. **C.** ability to lower pH.
D. ability to hydrogen bond. **E.** high melting points.

Questions 14 through17 refer to the following reactions:

A. $CH_3-\overset{O}{\overset{\|}{C}}-OH + CH_3-OH \xrightarrow{H^+} CH_3-\overset{O}{\overset{\|}{C}}-O-CH_3 + H_2O$

B. $CH_3-\overset{O}{\overset{\|}{C}}-OH + NaOH \longrightarrow CH_3-\overset{O}{\overset{\|}{C}}-O^-\,Na^+ + H_2O$

C. $CH_3-\overset{O}{\overset{\|}{C}}-O-CH_3 + H_2O \xrightarrow{H^+} CH_3-\overset{O}{\overset{\|}{C}}-OH + CH_3-OH$

D. $CH_3-\overset{O}{\overset{\|}{C}}-O-CH_3 + NaOH \longrightarrow CH_3-\overset{O}{\overset{\|}{C}}-O^-\,Na^+ + CH_3-OH$

14. _____ is an ester hydrolysis.

15. _____ is a neutralization

16. _____ is a saponification.

17. _____ is an esterification.

18. What is the name of the organic product of reaction A?

A. methyl acetate
B. acetic acid
C. methyl alcohol
D. acetaldehyde
E. ethyl methanoate

19. The compound with the highest boiling point is

A. formic acid.
B. acetic acid.
C. propanol.
D. propanoic acid..
E. ethyl acetate.

20. The ester produced from the reactions of 1-butanol and propanoic acid is

A. butyl propanoate.
B. butyl propanone.
C. propyl butyrate.
D. propyl butanone.
E. heptanoate.

21. The reaction of methyl acetate with NaOH produces

A. ethanol and formic acid
B. ethanol and sodium formate
C. ethanol and sodium ethanoate
D. methanol and acetic acid
E. methanol and sodium acetate

22. Identify the carboxylic acid and alcohol needed to produce

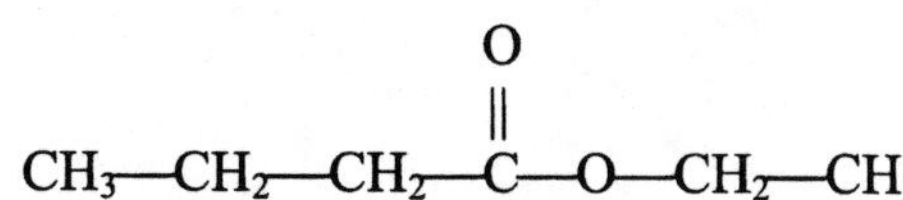

A. propanoic acid and ethanol
B. acetic acid and 1-pentanol
C. acetic acid and 1-butanol
D. butanoic acid and ethanol
E. hexanoic acid and methanol

23. When butanal is oxidized, the product is

A. butanone. **B.** 1-butanol. **C.** 2-butanol.
D. butanoic acid. **E.** butane.

24. The name of $CH_3—CH_2—\overset{\overset{\displaystyle O}{\|}}{C}—O—CH_2—CH_3$ is

A. ethyl acetate. **B.** ethyl ethanoate. **C.** ethyl propanoate.
D. propyl ethanoate. **E.** ethyl butyrate.

25. Soaps are

A. long chain fatty acids. **B.** fatty acid salts.
C. esters of acetic acid. **D.** alcohols with 10 carbon atoms.
E. aromatic compounds.

26. In a hydrolysis reaction,

A. an acid reacts with an alcohol.
B. an ester reacts with NaOH.
C. an ester reacts with H_2O.
D. an acid neutralizes a base.
E. water is added to an alkene.

27. Esters

A. have pleasant odors.
B. can undergo hydrolysis.
C. are formed from alcohols and carboxylic acids.
D. have a lower boiling point than the corresponding acid.
E. all of the above

28. The products of $H—\overset{\overset{\displaystyle O}{\|}}{C}—OH + H_2O$ are

A. $H—\overset{\overset{\displaystyle O}{\|}}{C}—O^- + H_3O^+$ **B.** $H—\overset{\overset{\displaystyle O}{\|}}{C}—\overset{+}{O}H_2 + OH^-$

C. $H—\overset{\overset{\displaystyle O}{\|}}{C}—O—CH_3$ **D.** $CH_3—\overset{\overset{\displaystyle O}{\|}}{C}—OH$

E. $H—\overset{\overset{\displaystyle O}{\|}}{C}—O^- Na^+ + H_2O$

29. The name of this compound is

[structure: benzene ring with $\overset{\overset{\displaystyle O}{\|}}{C}—OH$ substituent and Cl in the meta position]

A. *p*-chlorobenzoic acid. **B.** chlorobenzoic acid.
C. *m*-chlorobenzoic acid. **D.** 4-chlorobenzoic acid.
E. benzoic acid chloride.

30. The name of this compound is

O
||
C—O⁻ Na⁺ (attached to a benzene ring)

A. benzene sodium. **B.** sodium benzoate. **C.** *o*-benzoic acid.
D. sodium benzene carboxylate. **E.** benzoic acid.

Answers to the Practice Test

1. A	**2.** B	**3.** E	**4.** C	**5.** D
6. D	**7.** E	**8.** B	**9.** A	**10.** C
11. C	**12.** A	**13.** D	**14.** C	**15.** B
16. D	**17.** A	**18.** A	**19.** D	**20.** A
21. E	**22.** D	**23.** D	**24.** C	**25.** B
26. C	**27.** E	**28.** A	**29.** C	**30.** B

Answers and Solutions To Selected Text Problems

17.1 Methanoic acid (formic acid) is the carboxylic acid that is responsible for the pain associated with ant stings.

17.3 Each compound contains three carbon atoms. They differ because propanal, an aldehyde, contains a carbonyl group bonded to a hydrogen. In propanoic acid, the carbonyl group connects to a hydroxyl group.

17.5 **a.** Ethanoic acid (acetic acid) is the carboxylic acid with two carbons.
b. Butanoic acid (butyric acid) is the carboxylic acid with four carbons.
c. 2-chloropropanoic acid (α-chloropropionionic acid) is a three-carbon carboxylic acid with a chlorine on the carbon next to the carbonyl.
d. 3-methylhexanoic acid is a six-carbon carboxylic acid with a methyl on carbon 3.
e. 3,4-dihydroxybenzoicacid has a carboxylic acid group on benzene and two hydroxyl group on carbons 3, and 4.
f. 4-bromopentanoic acid is a five-carbon carboxylic acid with a –Br atom on carbon 4.

17.7

a. $CH_3—CH_2—C(=O)—OH$ Propionic acid has three carbons.

b. C(=O)—OH

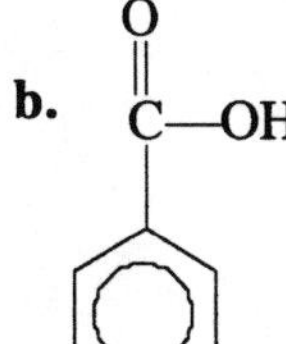

Benzoic acid is the carboxylic acid of benzene.

c. $Cl—CH_2—C(=O)—OH$ 2-chloroethanoic acid is a carboxylic acid that has a two-carbon chain with a chlorine atom on carbon 2.

d. $HO—CH_2—CH_2—C(=O)—OH$ 3-hydroxypropanoic acid is a carboxylic acid that has a three-carbon chain with a hydroxyl on carbon 3.

e. $CH_3-CH_2-\underset{\displaystyle CH_3}{CH}-\overset{\displaystyle O}{\overset{\|}{C}}-OH$ α-methylbutyric acid is a carboxylic acid that has a four-carbon chain with a methyl on the second (α) carbon.

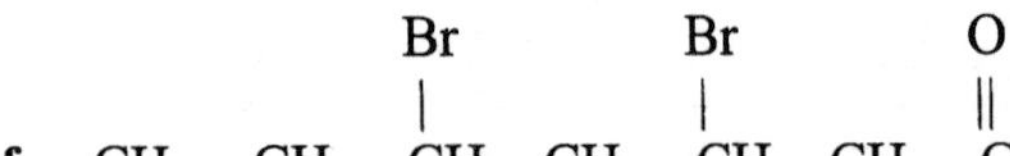

f. $CH_3-CH_2-\overset{\displaystyle Br}{CH}-CH_2-\overset{\displaystyle Br}{CH}-CH_2-\overset{\displaystyle O}{\overset{\|}{C}}-OH$ 3,5-dibromoheptanoic acid is a carboxylic acid that has a seven-carbon chain with two bromine atoms, one on carbon 3 and the other on carbon 5.

17.9 Aldehydes and primary alcohols oxidize to produce the corresponding carboxylic acid.

a. $H-\overset{\displaystyle O}{\overset{\|}{C}}-OH$

b. $CH_3-\overset{\displaystyle O}{\overset{\|}{C}}-OH$

c. $CH_3-\overset{\displaystyle CH_3}{\overset{|}{CH}}-CH_2-\overset{\displaystyle O}{\overset{\|}{C}}-OH$

d. cyclopentyl$-CH_2-\overset{\displaystyle O}{\overset{\|}{C}}-OH$

17.11
a. Butanoic acid has a higher molar mass and would have a higher boiling point.
b. Propanoic acid can form more hydrogen bonds and would have a higher boiling point.
c. Butanoic acid can form more hydrogen bonds and would have a higher boiling point.

17.13
a. acetone, propanol, propanoic acid. Propanoic acid forms the most hydrogen bonds of the three compounds.
b. butanoic acid, propanoic acid, acetic acid. Acetic acid has only two carbon atoms and the lowest molar mass of the group.
c. propane, ethanol, acetic acid. Acetic acid forms the most hydrogen bonds of the three compounds.

17.15
a. $H-\overset{\displaystyle O}{\overset{\|}{C}}-OH + H_2O \rightleftarrows H-\overset{\displaystyle O}{\overset{\|}{C}}-O^- + H_3O^+$

b. $CH_3-CH_2-\overset{\displaystyle O}{\overset{\|}{C}}-OH + H_2O \rightleftarrows CH_3-CH_2-\overset{\displaystyle O}{\overset{\|}{C}}-O^- + H_3O^+$

c. $CH_3-\overset{\displaystyle O}{\overset{\|}{C}}-OH + H_2O \leftharpoondown CH_3-\overset{\displaystyle O}{\overset{\|}{C}}-O^- + H_3O^+$

17.17 **a.** $H-\overset{O}{\overset{\|}{C}}-OH + NaOH \longrightarrow H-\overset{O}{\overset{\|}{C}}-O^- Na^+ + H_2O$

b. $CH_3-CH_2-\overset{O}{\overset{\|}{C}}-OH + NaOH \longrightarrow CH_3-CH_2-\overset{O}{\overset{\|}{C}}-O^- Na^+ + H_2O$

c. $C_6H_5-\overset{O}{\overset{\|}{C}}-OH$ + NaOH ⟶ $C_6H_5-\overset{O}{\overset{\|}{C}}-O^- Na^+$ + H_2O

17.19 A carboxylic acid salt is named by replacing the *-oic ic* ending of the acid name with *ate.*

a. The acid is methanoic acid (formic acid). The carboxylic acid salt is sodium methanoate, (sodium formate).
b. The acid is propanoic acid (propionic acid). The carboxylic acid salt is sodium propanoate (sodium propionate).
c. The acid is benzoic acid. The carboxylic acid salt is sodium benzoate.

17.21 **a.** This is an *aldehyde* since it has a carbonyl bonded to carbon and hydrogen.
b. This is an *ester* since it has a carbonyl bonded to oxygen that is also bonded to a carbon.
c. This is a *ketone* since it has a carbonyl bonded to two carbon atoms.
d. This is a *carboxylic acid* since it has a carboxylic group; a carbonyl bonded to a hydroxyl.

17.23 **a.** $CH_3-\overset{O}{\overset{\|}{C}}-O-CH_3$ The carbonyl portion of the ester has two carbons bonded to a one-carbon methyl group.

b. $CH_3-CH_2-CH_2-\overset{O}{\overset{\|}{C}}-OCH_3$ The carbonyl portion of the ester is a four-carbon chain bonded to a one-carbon methyl group.

c.

$C_6H_5-\overset{O}{\overset{\|}{C}}-O-CH_3$

17.25 A carboxylic acid and an alcohol react to give an ester with the elimination of water.

a. $CH_3-CH_2-\overset{O}{\overset{\|}{C}}-O-CH_2-CH_2-CH_3$

b. $CH_3-CH_2-CH_2-CH_2-\overset{O}{\overset{\|}{C}}-O-\overset{CH_3}{\overset{|}{CH}}-CH_3$

17.27 **a.** The carbonyl portion of the ester is derived from methanoic acid (formic acid). The alcohol is methanol (methyl alcohol).
b. The carbonyl portion of the ester is derived from ethanoic acid (acetic acid). The alcohol is methanol (methyl alcohol).

c. The carbonyl portion of the ester is derived from butanoic acid (butyric acid). The alcohol is methanol (methyl alcohol).

d. The carbonyl portion of the ester is derived from 3-methylbutanoic acid (β-methylbutyric acid). The alcohol is ethanol (ethyl alcohol).

17.29 **a.** The name of this ester is methyl methanoate (methyl formate). The carbonyl portion of the ester contains one carbon, the name is derived from methanoic (formic) acid. The alkyl portion has one carbon; it is methyl.

b. The name of this ester is methyl ethanoate (methyl acetate). The carbonyl portion of the ester contains two carbons; the name is derived from ethanoic (acetic) acid. The alkyl portion has one carbon; it is methyl.

c. The name of this ester is methyl butanoate (methyl butyrate). The carbonyl portion of the ester contains four carbons; the name is derived from butanoic (butyric) acid. The alkyl portion has one carbon; it is methyl.

d. The name of this ester is ethyl-3-methyl butanoate (ethyl-β-methyl butyrate). The carbonyl portion of the ester has a four-carbon chain with a methyl group attached to the third (β) carbon, counting the carboxyl carbon as 1. The alkyl portion with two carbons is an ethyl.

17.31 **a.** $CH_3—\overset{\overset{\displaystyle O}{\|}}{C}—O—CH_3$ Acetic acid is the two-carbon carboxylic acid. Methanol gives a one-carbon alkyl group.

b. $H—\overset{\overset{\displaystyle O}{\|}}{C}—O—CH_2—CH_2—CH_2—CH_3$ Formic acid is the carboxylic acid bonded to the four-carbon 1-butanol.

c. $CH_3—CH_2—CH_2—CH_2—\overset{\overset{\displaystyle O}{\|}}{C}—O—CH_2—CH_3$ Pentanoic acid is the carboxylic acid bonded to ethanol.

d. $CH3—CH2—\overset{\overset{\displaystyle O}{\|}}{C}—O—CH2—\overset{\overset{\displaystyle Br}{|}}{CH}—CH3$ Propanoic acid is the carboxylic acid bonded to

17.33 **a.** The flavor and odor of bananas is pentyl ethanoate (pentyl acetate).

b. The flavor and odor of oranges is octyl ethanoate (octyl acetate).

c. The flavor and odor of apricots is pentyl butanoate (pentyl butyrate).

d. The flavor and odor of raspberries is isobutyl methanoate (isobutyl formate).

17.35 **a.** $CH_3—\overset{\overset{\displaystyle O}{\|}}{C}—OH$ **b.** $CH_3—CH_2—CH_2—CH_2—OH$

c. $CH_3—O—\overset{\overset{\displaystyle O}{\|}}{C}—CH_3$

17.37 Acid hydrolysis of an ester adds water in the presence of acid and gives a carboxylic acid and an alcohol.

17.39 Acid hydrolysis of an ester gives the carboxylic acid and the alcohol, which were combined to form the ester; basic hydrolysis of an ester gives the salt of carboxylic acid and the alcohol, which combine to form the ester.

a. $CH_3{-}CH_2{-}\overset{\displaystyle O}{\overset{\|}{C}}{-}O^-\ Na^+$ and $CH_3{-}OH$

b. $CH_3{-}\overset{\displaystyle O}{\overset{\|}{C}}{-}OH$ and $CH_3{-}CH_2{-}CH_2{-}OH$

c. $CH_3{-}CH_2{-}CH_2{-}\overset{\displaystyle O}{\overset{\|}{C}}{-}OH$ and $CH_3{-}CH_2{-}OH$

d. $C_6H_5{-}COOH$ and CH_3CH_2OH

e. $C_6H_5{-}COO^-\ Na^+$ and CH_3CH_2OH

17.41 **a.** 3-methylbutanoic acid; *β*-methylbuyric acid
b. ethylbenzoate
c. ethyl propanoate; ethylpropionate
d. 2-chlorobenzoic acid; *ortho*chlorobenzoic acid
e. 4-hydroxypentanoic acid
f. 2-propyl ethanoate; isopropyl acetate

17.43 $CH_3{-}CH_2{-}CH_2{-}CH_2{-}\overset{\displaystyle O}{\overset{\|}{C}}{-}OH$ $\quad$ $CH_3{-}CH_2{-}\overset{\displaystyle CH_3}{\overset{|}{C}H}{-}\overset{\displaystyle O}{\overset{\|}{C}}{-}OH$

$CH_3{-}\overset{\displaystyle CH_3}{\overset{|}{C}H}{-}CH_2{-}\overset{\displaystyle O}{\overset{\|}{C}}{-}OH$ $\quad$ $CH_3{-}\underset{\displaystyle CH_3}{\underset{|}{\overset{\displaystyle CH_3}{\overset{|}{C}}}}{-}\overset{\displaystyle O}{\overset{\|}{C}}{-}OH$

17.45 **a.** $CH_3{-}O{-}\overset{\displaystyle O}{\overset{\|}{C}}{-}CH_3$ $\quad$ **b.** 4-chlorobenzoic acid ($COOH$ and Cl on a benzene ring, para)

c. $Cl{-}CH_2{-}CH_2{-}\overset{\displaystyle O}{\overset{\|}{C}}{-}OH$ $\quad$ **d.** $CH_3{-}CH_2{-}O{-}\overset{\displaystyle O}{\overset{\|}{C}}{-}CH_2{-}CH_2{-}CH_3$

e. $CH_3—CH_2—CH(CH_3)—CH_2—C(=O)—OH$ f. 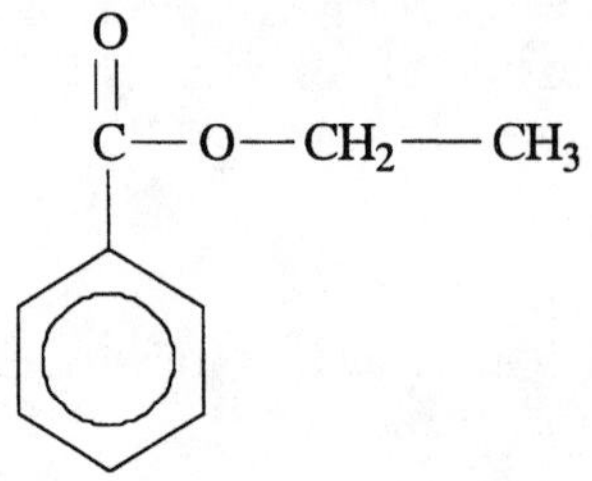 $C_6H_5—C(=O)—O—CH_2—CH_3$

17.47 a. $CH_3—C(=O)—OH$

b. $CH_3—CH_2—C(=O)—OH$

c. $CH_3—CH_2—CH_2—C(=O)—OH$

17.49 The presence of two polar groups in the carboxyl group allows hydrogen bonding, including the formation of a dimer that doubles the effective molar mass and requires a higher temperature to form gas.

17.51 b, c, d, and e are all soluble in water

17.53

$$C_6H_5—C(=O)—OCH_3 + KOH \longrightarrow C_6H_5—C(=O)—O^- K^+ + CH_3OH$$

A soluble salt, potassium benzoate, is formed. When acid is added, the salt is converted to insoluble benzoic acid.

17.55 a. hydroxyl and carboxylic acid

b. $C_6H_4(—O—C(=O)—CH_3)(—C(=O)—OH)$

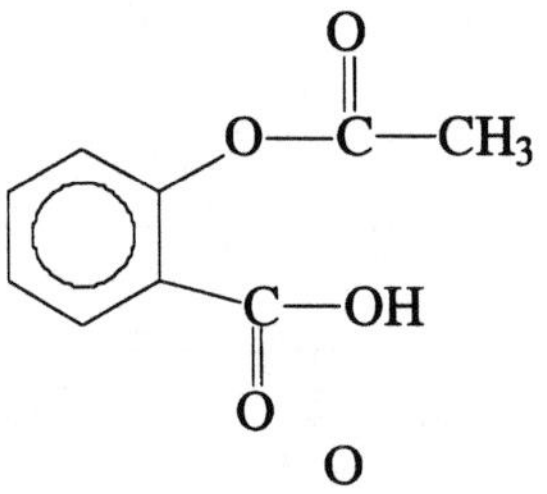

c. $C_6H_4(—OH)(—C(=O)—OCH_3)$

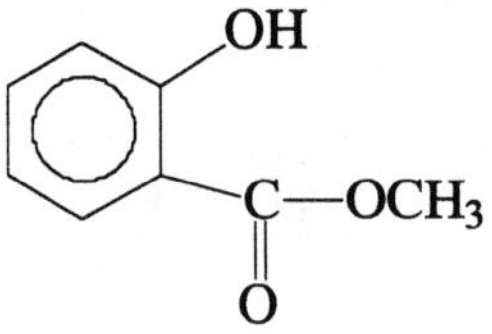

17.57 a. $CH_3—CH_2—C(=O)—O^- + H_3O^+$

b. $CH_3—CH_2—C(=O)—O^- K^+ + H_2O$

c. $CH_3—CH_2—C(=O)—O—CH_3 + H_2O$

d. $C_6H_5-\overset{O}{\overset{\|}{C}}-O-CH_3$

17.59 **a.** The acid 3-methylbutanoic acid is needed to react with methanol (CH_3-OH).
b. The acid 3-chlorobenzoic acid is needed to react with ethanol (CH_3-CH_2-OH).
c. The acid hexanoic acid is needed to react with methanol(CH_3-OH).

17.61 **a.** $CH_3-CH_2-\overset{O}{\overset{\|}{C}}-OH$ and $HO-\overset{CH_3}{\overset{|}{CH}}-CH_3$

b. $CH_3-\overset{CH_3}{\overset{|}{CH}}-\overset{O}{\overset{\|}{C}}-O^-\ Na^+$ and $HO-CH_2-CH_2-CH_3$

17.63 **a.** $CH_2{=}CH_2 + H_2O \xrightarrow{H^+} CH_3-CH_2-OH \xrightarrow{[O]} CH_3-\overset{O}{\overset{\|}{C}}-OH$

b. $CH_3-CH_2-CH_2-CH_2-OH \xrightarrow{[O]} CH_3-CH_2-CH_2-\overset{O}{\overset{\|}{C}}-OH$

18
Lipids

Study Goals

- Describe the properties and types of lipids.
- Write the structures of triacylglycerols obtained from glycerol and fatty acids.
- Draw the structure of the product from hydrogenation, hydrolysis, and saponification of triacylglycerols.
- Distinguish between phospholipids, glycolipids, and sphingolipids.
- Describe steroids and their role in bile salts, vitamins, and hormones.
- Describe the lipid bilayer in a cell.

Think About It

1. What are fats used for in the body?

2. What foods are high in fat?

3. What oils are used to produce margarines?

4. What kind of lipid is cholesterol?

Key Terms

Match the key terms with the correct statement shown below.

a. lipid **b.** fatty acid **c.** triacylglycerol
d. saponification **e.** phospholipid **f.** steroid

1. _____ A lipid consisting of glycerol bonded to two fatty acids and a phosphate group attached to an amino group

2. _____ A type of compound that is not soluble in water but is soluble in nonpolar solvents

3. _____ The hydrolysis of a triacylglycerol with a strong base producing salts called soaps and glycerol

4. _____ A lipid consisting of glycerol bonded to three fatty acids

5. _____ A lipid composed of a multicyclic ring system

6. _____ Long-chain carboxylic acid found in triacylglycerols

Answers **1.** e **2.** a **3.** d **4.** c **5.** f **6.** b

18.1 Lipids

- Lipids are nonpolar compounds that are not soluble in water.
- Classes of lipids include waxes, triacylglycerols, glycerophospholipids, and steroids.

◆ Learning Exercise 18.1

Match one of the classes of lipids with the composition of lipids below:

a. wax **b.** triacylglycerol **c.** glycerophospholipid
d. sphingolipid **e.** glycosphingolipid **f.** steroid

1. ____ A fused structure of four cycloalkanes.

2. ____ A long chain alcohol and a fatty acid

3. ____ Glycerol and three fatty acids

4. ____ Glycerol, two fatty acids, phosphate and choline

5. ____ Sphingosine, fatty acid, and galactose

6. ____ Sphingosine, fatty acid, phosphate and choline

Answers **1.** f **2.** a **3.** b **4.** c **5.** e **6.** d

18.2 Fatty Acids

- Fatty acids are unbranched carboxylic acids that typically contain an even number (12–18) of carbon atoms.
- Fatty acids may be saturated, monounsaturated with one double bond, or polyunsaturated with two or more double bonds. The double bonds in unsaturated fatty acids are almost always cis.

◆ Learning Exercise 18.2

Draw the structural formulas of linoleic acid, stearic acid, and oleic acid.

A. linoleic acid

B. stearic acid

C. oleic acid

Which of these three fatty acids

1. ____ is the most saturated? **2.** ____ is the most unsaturated?

3. ____ has the lowest melting point? **4.** ____ has the highest melting point?

5. ____ is found in vegetables? **6.** ____ is from animal sources?

Answers

linoleic acid $CH_3—(CH_2)_4—CH{=}CH—CH_2—CH{=}CH—(CH_2)_7—\overset{\overset{\displaystyle O}{\|}}{C}—OH$

stearic acid $CH_3—(CH_2)_{16}—\overset{\overset{\displaystyle O}{\|}}{C}—OH$

oleic acid $CH_3—(CH_2)_7—CH{=}CH—(CH_2)_7—\overset{\overset{\displaystyle O}{\|}}{C}—OH$

1. B **2.** A **3.** A **4.** B **5.** A and C **6.** B

18.3 Waxes, Fats, and Oils

- A wax is an ester of a long-chain fatty acid and a long-chain alcohol.
- The triacylglycerols in fats and oils are esters of glycerol with three long-chain fatty acids.
- Fats from animal sources contain more saturated fatty acids and have higher melting points than fats found in most vegetable oils.

◆ Learning Exercise 18.3A

Write the formula of the wax formed by the reaction of palmitic acid, $CH_3—(CH_2)_{14}—COOH$, and cetyl alcohol, $CH_3—(CH_2)_{14}—CH_2—OH$.

Answer $CH_3—(CH_2)_{14}—\overset{\overset{\displaystyle O}{\|}}{C}—O—CH_2—(CH_2)_{14}—CH_3$

◆ Learning Exercise 18.3B

Consider the following fatty acid called oleic acid.

[skeletal structure of oleic acid with C=O and OH]

O

OH

1. Why is the compound an acid?

2. Is it a saturated or unsaturated compound? Why?

3. Is the double bond cis or trans?

4. Is it likely to be a solid or a liquid at room temperature?

5. Why is it not soluble in water?

Answers

1. contains a carboxylic acid group
2. unsaturated; double bond
3. cis
4. liquid
5. it has a long hydrocarbon chain

◆ Learning Exercise 18.3C

Write the structure and name of the triacylglycerol formed from the following.

1. Glycerol and three palmitic acids, $CH_3—(CH_2)_{14}—COOH$

2. Glycerol and three myristic acids, $CH_3—(CH_2)_{12}—COOH$

Answers

1.

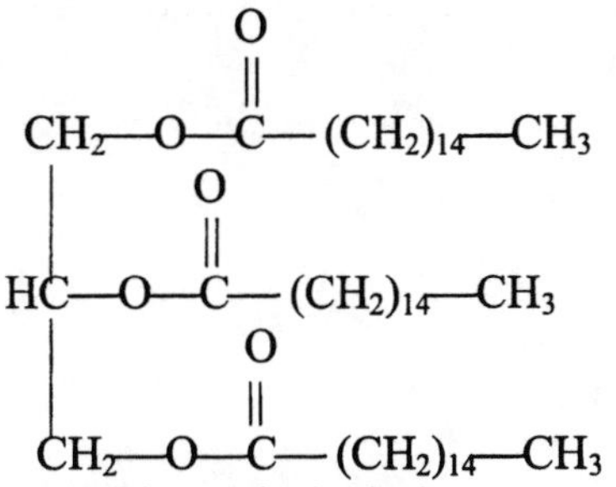

$$\begin{array}{l} CH_2—O—\overset{\overset{\displaystyle O}{\|}}{C}—(CH_2)_{14}—CH_3 \\ | \\ HC—O—\overset{\overset{\displaystyle O}{\|}}{C}—(CH_2)_{14}—CH_3 \\ | \\ CH_2—O—\overset{\overset{\displaystyle O}{\|}}{C}—(CH_2)_{14}—CH_3 \end{array}$$

Glyceryl tripalmitate
(Tripalmitin)

2.

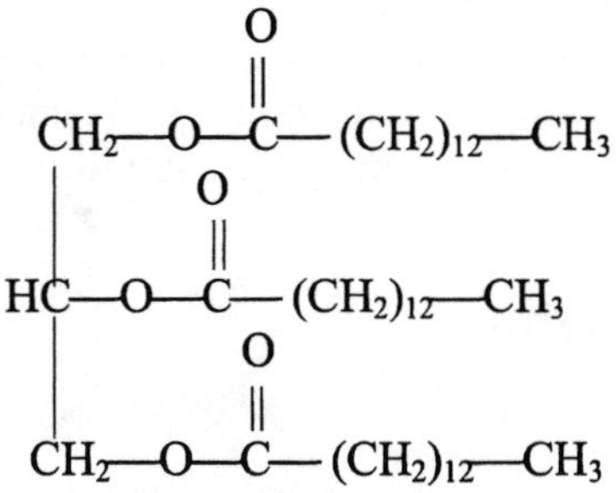

$$\begin{array}{l} CH_2—O—\overset{\overset{\displaystyle O}{\|}}{C}—(CH_2)_{12}—CH_3 \\ | \\ HC—O—\overset{\overset{\displaystyle O}{\|}}{C}—(CH_2)_{12}—CH_3 \\ | \\ CH_2—O—\overset{\overset{\displaystyle O}{\|}}{C}—(CH_2)_{12}—CH_3 \end{array}$$

Glyceryl trimyristate
(Trimyristin)

◆ Learning Exercise 18.3D

Write the structural formulas of the following triacylglycerols:

1. Glyceryl tristearate (Tristearin)

2. Glyceryl trioleate (Triolein)

Answers

1.

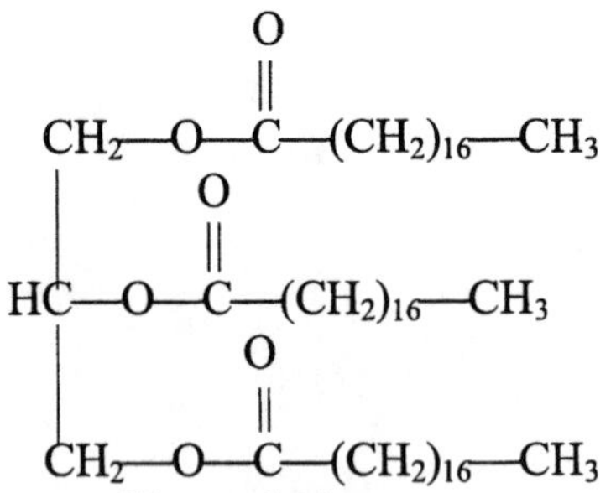

$$\begin{array}{l} \mathrm{CH_2{-}O{-}\overset{\displaystyle O}{\overset{\|}{C}}{-}(CH_2)_{16}{-}CH_3} \\ \mathrm{HC{-}O{-}\overset{\displaystyle O}{\overset{\|}{C}}{-}(CH_2)_{16}{-}CH_3} \\ \mathrm{CH_2{-}O{-}\overset{\displaystyle O}{\overset{\|}{C}}{-}(CH_2)_{16}{-}CH_3} \end{array}$$

Glyceryl tristearate
(Tristearin)

2.

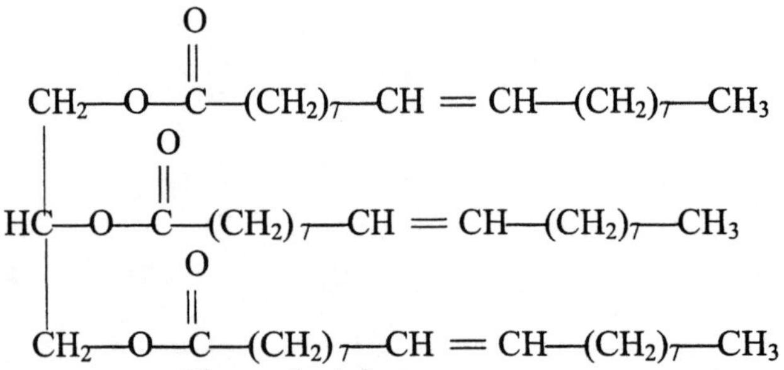

$$\begin{array}{l} \mathrm{CH_2{-}O{-}\overset{\displaystyle O}{\overset{\|}{C}}{-}(CH_2)_7{-}CH = CH{-}(CH_2)_7{-}CH_3} \\ \mathrm{HC{-}O{-}\overset{\displaystyle O}{\overset{\|}{C}}{-}(CH_2)_7{-}CH = CH{-}(CH_2)_7{-}CH_3} \\ \mathrm{CH_2{-}O{-}\overset{\displaystyle O}{\overset{\|}{C}}{-}(CH_2)_7{-}CH = CH{-}(CH_2)_7{-}CH_3} \end{array}$$

Glyceryl trioleate
(Triolein)

18.4 Chemical Properties of Triacylglycerols

- The hydrogenation of unsaturated fatty acids converts double bonds to single bonds.
- The oxidation of unsaturated fatty acids produces short-chain fatty acids with disagreeable odors.
- The hydrolysis of the ester bonds in fats or oils produces glycerol and fatty acids.
- In saponification, a fat heated with a strong base produces glycerol and the salts of the fatty acids or soaps. The dual polarity of soap permits its solubility in both water and oil.

◆ Learning Exercise 18.4

Write the equations for the following reactions of glyceryl trioleate (triolein)

1. hydrogenation with a nickel catalyst

2. acid hydrolysis with HCl

3. saponification with NaOH

Answers

1.

$$\begin{array}{l} CH_2-O-\overset{\overset{\displaystyle O}{\|}}{C}-(CH_2)_7-CH{=}CH-(CH_2)_7-CH_3 \\ | \\ HC-O-\overset{\overset{\displaystyle O}{\|}}{C}-(CH_2)_7-CH{=}CH-(CH_2)_7-CH_3 + 3H_2 \xrightarrow{Ni} \\ | \\ CH_2-O-\overset{\overset{\displaystyle O}{\|}}{C}-(CH_2)_7-CH{=}CH-(CH_2)_7-CH_3 \end{array} \quad \begin{array}{l} CH_2-O-\overset{\overset{\displaystyle O}{\|}}{C}-(CH_2)_{16}-CH_3 \\ | \\ HC-O-\overset{\overset{\displaystyle O}{\|}}{C}-(CH_2)_{16}-CH_3 \\ | \\ CH_2-O-\overset{\overset{\displaystyle O}{\|}}{C}-(CH_2)_{16}-CH_3 \end{array}$$

2.

$$\begin{array}{l} CH_2-O-\overset{\overset{\displaystyle O}{\|}}{C}-(CH_2)_7-CH{=}CH-(CH_2)_7-CH_3 \\ | \\ HC-O-\overset{\overset{\displaystyle O}{\|}}{C}-(CH_2)_7-CH{=}CH-(CH_2)_7-CH_3 + 3H_2O \xrightarrow{H^+} \\ | \\ CH_2-O-\overset{\overset{\displaystyle O}{\|}}{C}-(CH_2)_7-CH{=}CH-(CH_2)_7-CH_3 \end{array} \quad \begin{array}{l} CH_2-OH \\ | \\ HC-OH \\ | \\ CH_2-OH \end{array}$$

$$+\ 3\ HO-\overset{\overset{\displaystyle O}{\|}}{C}-(CH_2)_7-CH{=}CH-(CH_2)_7-CH_3$$

3.

$$\begin{array}{l} CH_2-O-\overset{\overset{\displaystyle O}{\|}}{C}-(CH_2)_7-CH{=}CH-(CH_2)_7-CH_3 \\ | \\ HC-O-\overset{\overset{\displaystyle O}{\|}}{C}-(CH_2)_7-CH{=}CH-(CH_2)_7-CH_3 + 3NaOH \longrightarrow \\ | \\ CH_2-O-\overset{\overset{\displaystyle O}{\|}}{C}-(CH_2)_7-CH{=}CH-(CH_2)_7-CH_3 \end{array} \quad \begin{array}{l} CH_2-OH \\ | \\ HC-OH \\ | \\ CH_2-OH \end{array}$$

$$+\ 3\ Na^+\ {}^-O-\overset{\overset{\displaystyle O}{\|}}{C}-(CH_2)_7-CH{=}CH-(CH_2)_7-CH_3$$

18.5 Glycerophospholipids

- Glycerophospholipids are esters of glycerol with two fatty acids and a phosphate group attached to an amino alcohol.
- The fatty acids are a nonpolar region, whereas the phosphate group and the amino alcohol make up a polar region.

◆ Learning Exercise 18.5A

Draw the structure of a glycerophospholipid that is formed from two molecules of palmitic acid and serine, an amino alcohol.

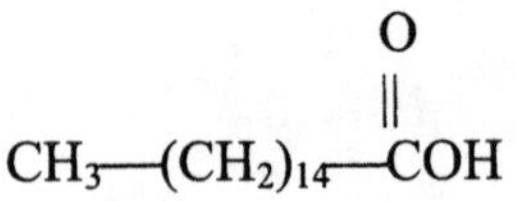

$$CH_3{-}(CH_2)_{14}{-}\overset{\overset{\displaystyle O}{\|}}{C}OH$$

Palmitic acid

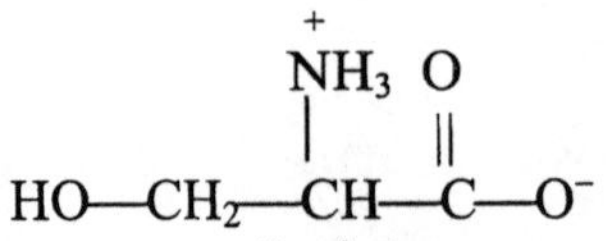

$$HO{-}CH_2{-}\overset{\overset{\displaystyle \overset{+}{N}H_3}{|}}{CH}{-}\overset{\overset{\displaystyle O}{\|}}{C}{-}O^-$$

Serine

Answer

$$\begin{array}{l} CH_2{-}O{-}\overset{\overset{\displaystyle O}{\|}}{C}{-}(CH_2)_{14}{-}CH_3 \\ | \\ HC{-}O{-}\overset{\overset{\displaystyle O}{\|}}{C}{-}(CH_2)_{14}{-}CH_3 \\ | \\ CH_2{-}O{-}\underset{\underset{\displaystyle O^-}{|}}{\overset{\overset{\displaystyle O}{\|}}{P}}{-}O{-}CH_2{-}\overset{\overset{\displaystyle \overset{+}{N}H_3}{|}}{CH}{-}\overset{\overset{\displaystyle O}{\|}}{C}{-}O^- \end{array}$$

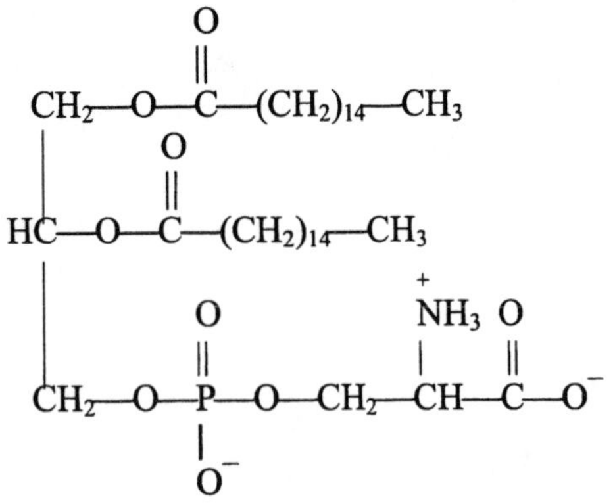

◆ Learning Exercise 18.5B

Consider the following glycerophospholipid.

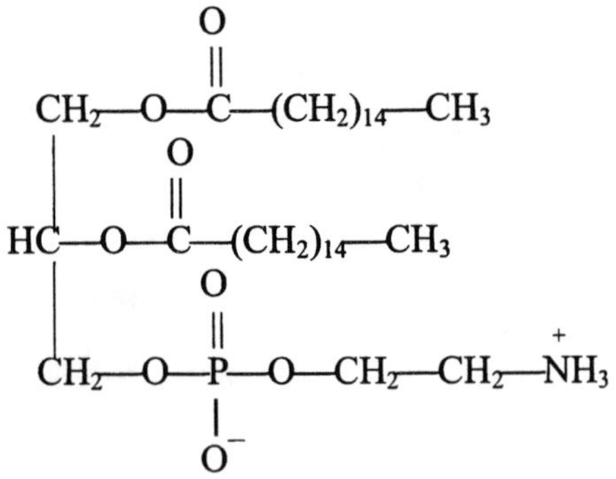

$$\begin{array}{l} CH_2{-}O{-}\overset{\overset{\displaystyle O}{\|}}{C}{-}(CH_2)_{14}{-}CH_3 \\ | \\ HC{-}O{-}\overset{\overset{\displaystyle O}{\|}}{C}{-}(CH_2)_{14}{-}CH_3 \\ | \\ CH_2{-}O{-}\underset{\underset{\displaystyle O^-}{|}}{\overset{\overset{\displaystyle O}{\|}}{P}}{-}O{-}CH_2{-}CH_2{-}\overset{+}{N}H_3 \end{array}$$

On the above structure, indicate the

A. two fatty acids.
B. the part from the glycerol molecule.
C. the phosphate section.
D. the amino alcohol group.
E. the nonpolar region.
F. the polar region.

1. What is the name of the amino alcohol group? ________________

2. What is the name of the phosphoglyceride? ________________

3. Why is a phosphoglyceride more soluble in water than most lipids? ________________

Answers

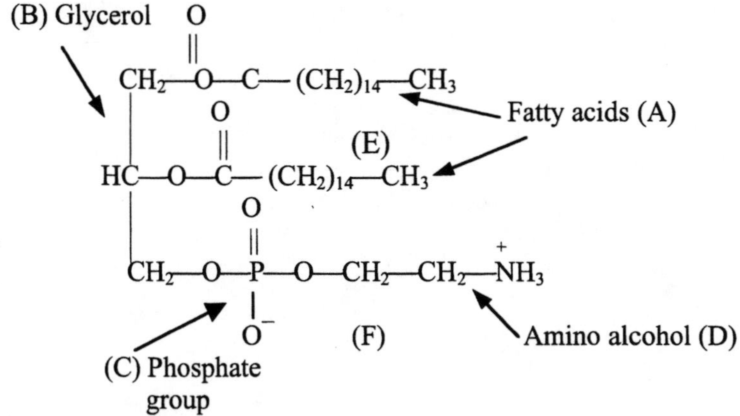

1. ethanolamine
2. ethanolamine phosphoglyceride
3. The polar portion of the phosphoglyceride is attracted to water, which makes this type of lipid more soluble in water than other lipids.

18.6 Sphingolipids

- In sphingolipids, the alcohol sphingosine forms an ester bond with one fatty acid and the phosphate-amino alcohol group.
- In glycosphingolipids, sphingosine is bonded to fatty acids and one or more monosaccharides.

◆ Learning Exercise 18.6

Match the following statements with one of the following types of phospholipids.

a. glycerophospholipid **b.** sphingolipid **c.** cerebroside **d.** ganglioside

1. _____ Contains sphingosine, a fatty acid and two or more monosaccharides

2. _____ Contains sphingosine, a fatty acid, phosphate, and an amino alcohol

3. _____ Contains sphingosine, a fatty acid, and one monosaccharide

4. _____ Contains glycerol, two fatty acids, phosphate, and an amino alcohol

Answers **1.** d **2.** b **3.** c **4.** a

18.7 Steroids: Cholesterol, Bile Salts, and Steroid Hormones

- Steroids are lipids containing the steroid nucleus, which is a fused structure of four rings.
- Steroids include cholesterol, bile salts, and vitamin D.
- The steroid hormones are closely related in structure to cholesterol and depend on cholesterol for their synthesis. The sex hormones such as estrogen and testosterone are responsible for sexual characteristics and reproduction. The adrenal corticosteroids such as aldosterone and cortisone regulate water balance and glucose levels in the cells.

◆ Learning Exercise 18.7A

1. Write the structure of the steroid nucleus.

2. Write the structure of cholesterol.

Answer

1.

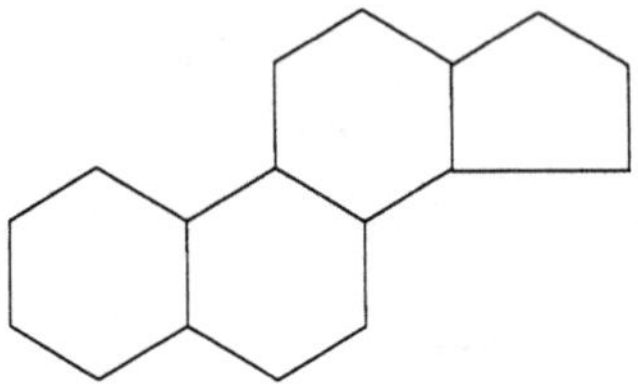

2.

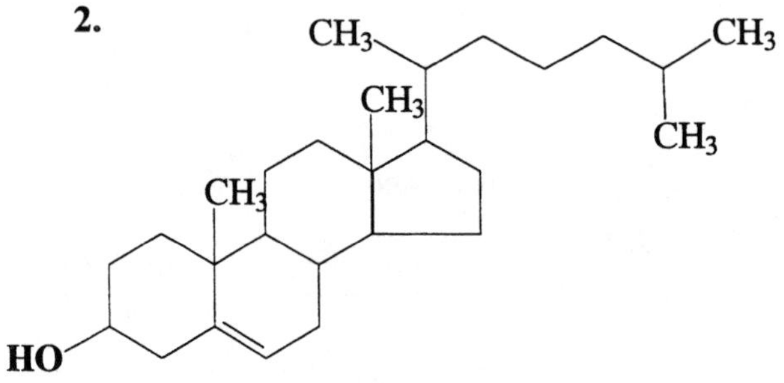

CH_3 CH_3 CH_3 CH_3 CH_3 HO

◆ Learning Exercise 18.7B

Identify one of these compounds with the following statements.

a. estrogen **b.** testosterone **c.** cortisone **d.** aldosterone **e.** bile salts

1. _____ Increases the blood level of glucose

2. _____ Increases the reabsorption of Na^+ in the blood

3. _____ Stimulates development of secondary sex characteristics in females

4. _____ Stimulates reabsorption of water by the kidneys

5. _____ Stimulates the secondary sex characteristics in males

6. _____ Secreted from the gallbladder into the small intestine to emulsify fats in the diet

Answers **1.** c **2.** d **3.** a **4.** d **5.** b **6.** e

18.8 Cell Membranes

- Cell membranes surround all of our cells and separate the cellular contents from the external liquid environment.
- A cell membrane is a lipid bilayer composed of two rows of phospholipids such that the nonpolar hydrocarbon tails are in the center and the polar sections are aligned along the outside.
- The inner portion of the lipid bilayer consists of nonpolar chains of the fatty acids, with the polar heads at the outer and inner surfaces.
- Molecules of cholesterol, proteins, glycolipids, and glycoproteins are embedded in the lipid bilayer.

◆ Learning Exercise 18.8

a. What is the function of the lipid bilayer in cell membranes?

b. What type of lipid makes up the lipid bilayer?

c. What is the general arrangement of the lipids in a lipid bilayer?

d. What are the functions of the proteins embedded in the lipid bilayer

Answers

a. The lipid bilayer separates the contents of a cell from the surrounding aqueous environment.
b. The lipid bilayer is composed of phospholipids.
c. The nonpolar hydrocarbon tails are in the center of the bilayer, while the polar sections are aligned along the outside of the bilayer.
d. Some proteins provide channels for electrolytes and water to flow in and out of the cell. Other proteins act as receptors for chemicals such as hormones, neurotransmitters, and antibiotics.

Check List for Chapter 18

You are ready to take the practice test for Chapter 18. Be sure that you have accomplished the following learning goals for this chapter. If you are not sure, review the section listed at the end of the goal. Then apply your new skills and understanding to the practice test.

After studying Chapter 18, I can successfully:

____ Describe the classes of lipids (18.1).

____ Identify a fatty acid as saturated or unsaturated (18.2).

____ Write the structural formula of a wax or triacylglycerol produced by the reaction of a fatty acid and an alcohol or glycerol (18.3).

____ Draw the structure of the product from the reaction of a triacylglycerol with hydrogen, an acid or base, or an oxidizing agent (18.4).

____ Describe the components of glycerolphospholipids (18.5).

____ Describe the components of sphingolipids and glycosphingolipids (18.6).

____ Describe the structure of a steroid and cholesterol (18.7).

____ Describe the function of the lipid bilayer in cell membranes (18.8).

Practice Test for Chapter 18

1. An ester of a fatty acid is called a

A. carbohydrate **B.** lipid **C.** protein **D.** oxyacid **E.** soap

2. A fatty acid that is unsaturated is usually from

A. animal sources and liquid at room temperature.
B. animal sources and solid at room temperature.
C. vegetable sources and liquid at room temperature.
D. vegetable sources and solid at room temperature.
E. both vegetable and animal sources and solid at room temperature.

3. 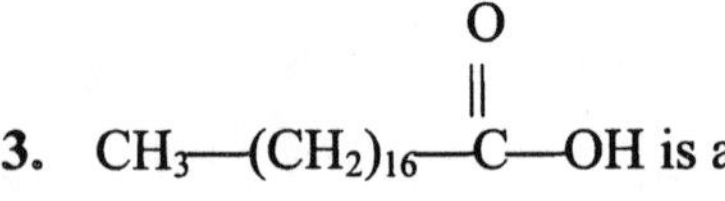 is a

A. unsaturated fatty acid **B.** saturated fatty acid **C.** wax
D. triacylglycerol **E.** sphingolipid

For questions 4 through 7, consider the following compound.

$$\begin{array}{l}
CH_2-O-\overset{\displaystyle O}{\overset{\|}{C}}-(CH_2)_{16}-CH_3 \\
| \\
HC-O-\overset{\displaystyle O}{\overset{\|}{C}}-(CH_2)_{16}-CH_3 \\
| \\
CH_2-O-\overset{\displaystyle O}{\overset{\|}{C}}-(CH_2)_{16}-CH_3
\end{array}$$

4. This compound belongs in the family called

A. wax.
B. triacylglycerol.
C. phosphoglyceride.
D. sphingolipid.
E. steroid.

5. The molecule shown above was formed by

A. esterification.
B. hydrolysis (acid).
C. saponification.
D. emulsification.
E. oxidation.

6. If this molecule were reacted with a strong base such as NaOH, the products would be

A. glycerol and fatty acids.
B. glycerol and water.
C. glycerol and soap.
D. an ester and salts of fatty acids.
E. an ester and fatty acids.

7. The compound would be expected to be

A. saturated, and a solid at room temperature.
B. saturated, and a liquid at room temperature.
C. unsaturated, and a solid at room temperature.
D. unsaturated, and a liquid at room temperature.
E. supersaturated, and a liquid at room temperature.

8. Which are found in glycerolphospholipids?

A. fatty acids
B. glycerol
C. a nitrogen compound
D. phosphate
E. all of these

For questions 9 and 10, consider the following reaction:

Triacylglycerol + 3NaOH $\longrightarrow$ 3 sodium salts of fatty acids and glycerol

9. The reaction of a triacylglycerol with a strong base such as NaOH is called

A. esterification.
B. lipogenesis.
C. hydrolysis.
D. saponification.
E. β-oxidation.

10. What is another name for the sodium salts of the fatty acids?

A. margarines
B. fat substitutes
C. soaps
D. perfumes
E. vitamins

For questions 11 through 16, consider the following phosphoglyceride.

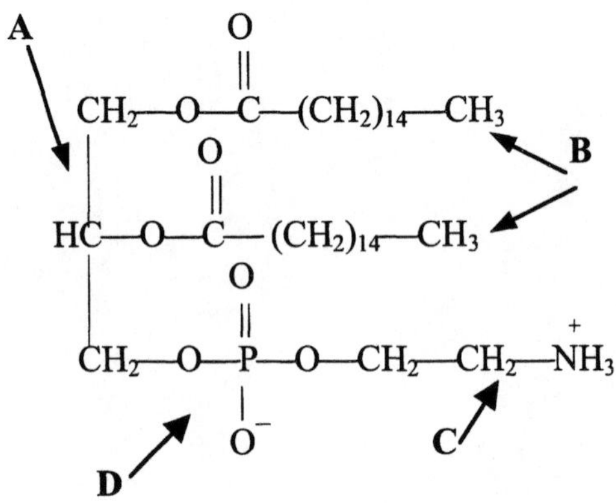

Match the labels in the phosphoglyceride with the following:

11. _____ The glycerol portion **12.** _____ The phosphate portion **13.** _____ The amino alcohol

14. _____ The polar region **15.** _____ The nonpolar region

16. Type of phospholipid

A. choline **B.** cephalin **C.** sphingomyelin
D. glycolipid **E.** cerebroside

Classify the lipids in 17 through 20 as
A. wax **B.** triacylglycerol **C.** phosphoglyceride
D. steroid **E.** fatty acid

17. _____ cholesterol

18. _____ $CH_3—(CH_2)_{14}—\overset{\overset{\large O}{||}}{C}—OH$

19. _____ $CH_3—(CH_2)_{14}—\overset{\overset{\large O}{||}}{C}—O—(CH_2)_{30}—CH_3$

20. _____ an ester of glycerol with three palmitic acids

Select answers for 21 through 25 from the following:

A. testosterone **B.** estrogen **C.** prednisone
D. cortisone **E.** aldosterone

21. _____ Stimulates the female sexual characteristics

22. _____ Increases the retention of water by the kidneys

23. _____ Stimulates the male sexual characteristics

24. _____ Increases the blood glucose level

25. _____ Used medically to reduce inflammation and treat asthma

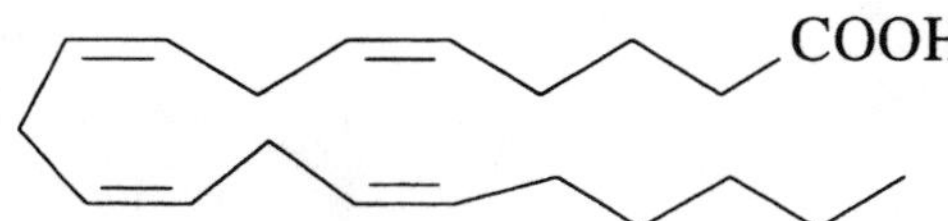

26. This compound is a

A. cholesterol. **B** sphingosine. **C.** prostaglandin.
D. glycerosphingolipid. **E.** steroid..

27. The lipid bilayer of a cell is composed of

A. cholesterol. **B** glycerophospholipids. **C.** proteins.
D. glycosphingolipids. **E.** all of these

28. The type of transport that carries chloride ions through integral proteins in the cell membrane is

A. passive transport. **B** active transport. **C.** diffusion.
D. facilitated transport. **E.** all of these

29. The movement of small molecules through a cell membrane from a higher concentration to a lower concentration is

A. passive transport. **B.** active transport. **C.** diffusion.
D. facilitated transport. **E.** A and C

30. The type of lipoprotein that transport cholesterol to the liver for elimination is called

A. chylomicron. **B.** high density lipoprotein **C.** low density lipoprotein.
D. very low density lipoprotein. **E.** all of these

Answers to the Practice Test

1. B	**2.** C	**3.** B	**4.** B	**5.**A
6. C	**7.** A	**8.** E	**9.** D	**10.** C
11. A	**12.** D	**13.** C	**14.** C, D	**15.** B
16. B	**17.** D	**18.** E	**19.** A	**20.** B
21. B	**22.** E	**23.** A	**24.** D	**25.** C
26. C	**27.** E	**28.** D	**29.** E	**30.** B

Answers and Solutions to Selected Text Problems

18.1 Lipids provide energy, protection, and insulation for the organs in the body.

18.3 Since lipids are not soluble in water, they are nonpolar molecules.

18.5. All fatty acids contain a long chain of carbon atoms with a carboxylic acid group. Saturated fats contain only carbon-to-carbon single bonds; unsaturated fats contain one or more double bonds. More saturated fats are from animal sources, while vegetable oils contain more unsaturated fatty acids.

18.7.

a. palmitic acid COOH

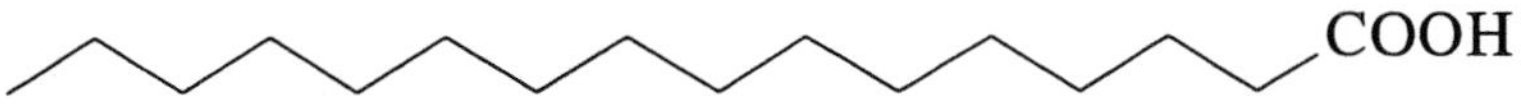

b. oleic aicd COOH

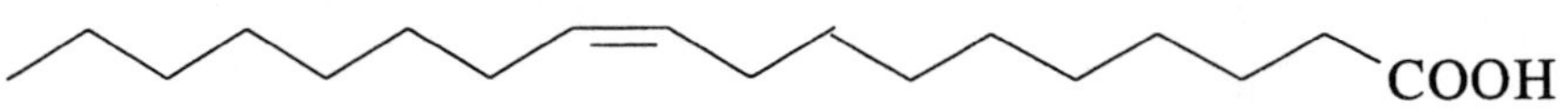

18.9 **a.** Lauric acid has only carbon–carbon single bonds; it is saturated.
b. Linolenic acid has three carbon–carbon double bonds; it is unsaturated.
c. Palmitoleic acid has one carbon–carbon double bond; it is unsaturated.
d. Stearic acid has only carbon–carbon single bonds; it is saturated.

18.11 In a cis double bond, the alkyl groups are on the same side of the double bond, whereas in trans fatty acids, the alkyl groups are on opposite sides.

18.13 In an omega-3 fatty acid, the first double bond occurs at carbon 3 counting from the methyl. In an omega-6 fatty acid, the first double bond occurs at carbon 6.

18.15 Arachidonic acid contains four double bonds and no side groups. In PGE_2, a part of the chain forms cyclopentane and there are hydroxyl and ketone functional groups.

18.17 Prostaglandins affect blood pressure, stimulate contraction and relaxation of smooth muscle.

18.19 Palmitic acid is the 16-carbon saturated fatty acid. $CH_3—(CH_2)_{14}—\overset{\overset{O}{\|}}{C}—O—(CH_2)_{29}—CH_3$

18.21 Fats are composed of fatty acids and glycerol. In this case, the fatty acid is stearic acid, an 18-carbon saturated fatty acid.

$$\begin{array}{l} CH_2—O—\overset{\overset{O}{\|}}{C}—(CH_2)_{16}—CH_3 \\ | \\ HC—O—\overset{\overset{O}{\|}}{C}—(CH_2)_{16}—CH_3 \\ | \\ CH_2—O—\overset{\overset{O}{\|}}{C}—(CH_2)_{16}—CH_3 \end{array}$$

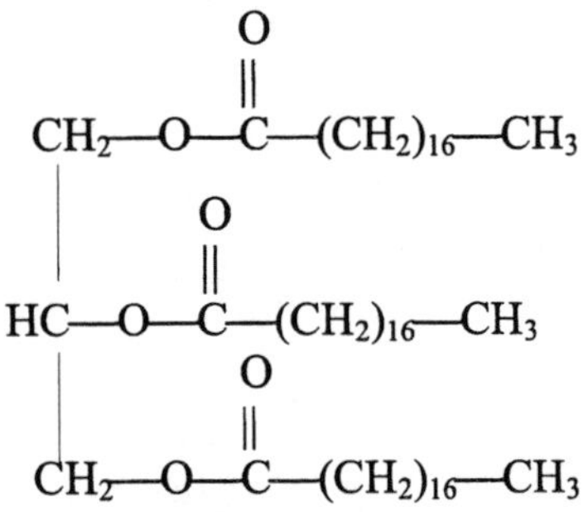

18.23 Tripalmitin has three palmitic acids (16-carbon saturated fatty acid) forming ester bonds with glycerol.

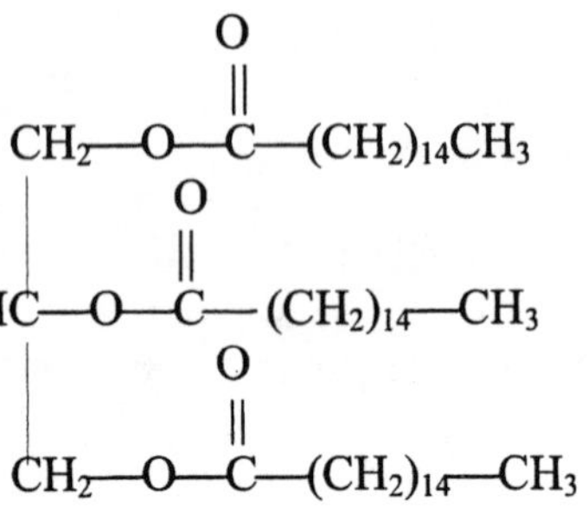

$$\begin{array}{l} CH_2—O—\overset{\overset{O}{\|}}{C}—(CH_2)_{14}CH_3 \\ | \\ HC—O—\overset{\overset{O}{\|}}{C}—(CH_2)_{14}—CH_3 \\ | \\ CH_2—O—\overset{\overset{O}{\|}}{C}—(CH_2)_{14}—CH_3 \end{array}$$

18.25 Safflower oil contains fatty acids with two or three double bonds; olive oil contains a large amount of oleic acid, which has a single (monounsaturated) double bond.

18.27 Although coconut oil comes from a vegetable source, it has large amounts of saturated fatty acids and small amounts of unsaturated fatty acids. Since coconut oil contains the same kinds of fatty acids as animal fat, coconut oil has a melting point similar to the melting point of animal fats.

18.29

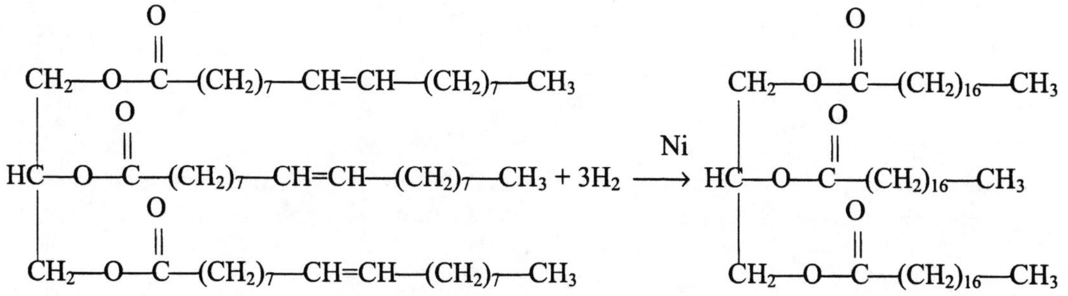

$$\begin{array}{l} CH_2-O-\overset{\displaystyle O}{\overset{\|}{C}}-(CH_2)_7-CH{=}CH-(CH_2)_7-CH_3 \\ HC-O-\overset{\displaystyle O}{\overset{\|}{C}}-(CH_2)_7-CH{=}CH-(CH_2)_7-CH_3 \\ CH_2-O-\overset{\displaystyle O}{\overset{\|}{C}}-(CH_2)_7-CH{=}CH-(CH_2)_7-CH_3 \end{array} + 3H_2 \xrightarrow{Ni} \begin{array}{l} CH_2-O-\overset{\displaystyle O}{\overset{\|}{C}}-(CH_2)_{16}-CH_3 \\ HC-O-\overset{\displaystyle O}{\overset{\|}{C}}-(CH_2)_{16}-CH_3 \\ CH_2-O-\overset{\displaystyle O}{\overset{\|}{C}}-(CH_2)_{16}-CH_3 \end{array}$$

18.31 **a.** Partial hydrogenation means that some of the double bonds in the unsaturated fatty acids have been converted to single bonds.

b. Since the margarine now has mostly saturated fatty acids, which can interact more strongly, it will be a solid.

18.33 Acid hydrolysis of a fat gives glycerol and the fatty acids. Basic hydrolysis (saponification) of fat gives glycerol and the salts of the fatty acids.

a.

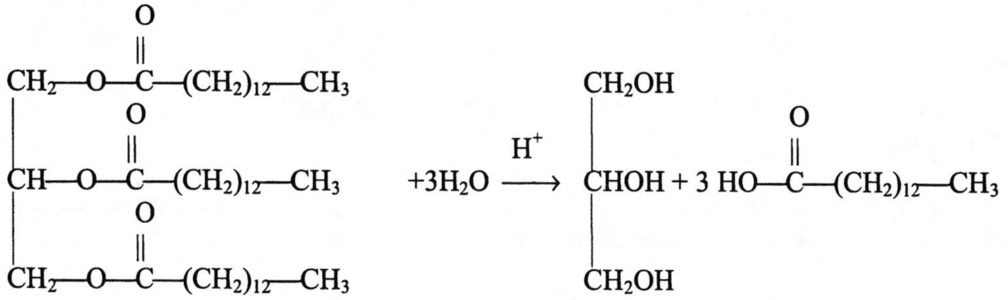

$$\begin{array}{l} CH_2-O-\overset{\displaystyle O}{\overset{\|}{C}}-(CH_2)_{12}-CH_3 \\ CH-O-\overset{\displaystyle O}{\overset{\|}{C}}-(CH_2)_{12}-CH_3 \\ CH_2-O-\overset{\displaystyle O}{\overset{\|}{C}}-(CH_2)_{12}-CH_3 \end{array} + 3H_2O \xrightarrow{H^+} \begin{array}{l} CH_2OH \\ CHOH \\ CH_2OH \end{array} + 3\ HO-\overset{\displaystyle O}{\overset{\|}{C}}-(CH_2)_{12}-CH_3$$

b.

$$\begin{array}{l} CH_2-O-\overset{\displaystyle O}{\overset{\|}{C}}-(CH_2)_{12}-CH_3 \\ CH-O-\overset{\displaystyle O}{\overset{\|}{C}}-(CH_2)_{12}-CH_3 \\ CH_2-O-\overset{\displaystyle O}{\overset{\|}{C}}-(CH_2)_{12}-CH_3 \end{array} + 3NaOH \longrightarrow \begin{array}{l} CH_2OH \\ CHOH \\ CH_2OH \end{array} + 3\ Na^{+}\,{}^{-}O-\overset{\displaystyle O}{\overset{\|}{C}}-(CH_2)_{12}-CH_3$$

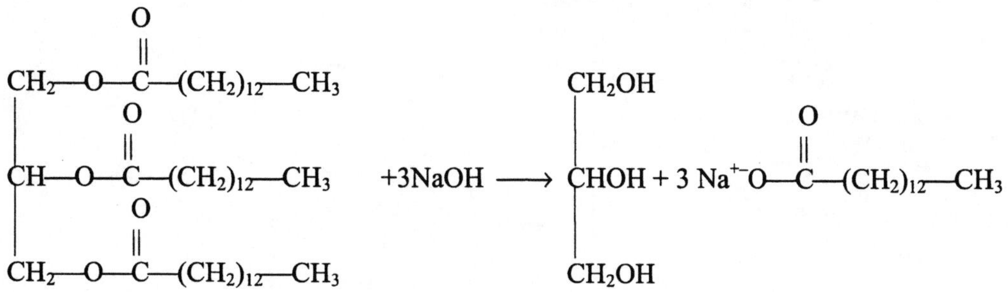

18.35 A triacylglycerol is a combination of three fatty acids bonded to glycerol by ester bonds. Olestra is sucrose bonded to six to eight fatty acids by ester bonds. Olestra cannot be digested because digestive enzymes cannot break down the molecule.

18.37

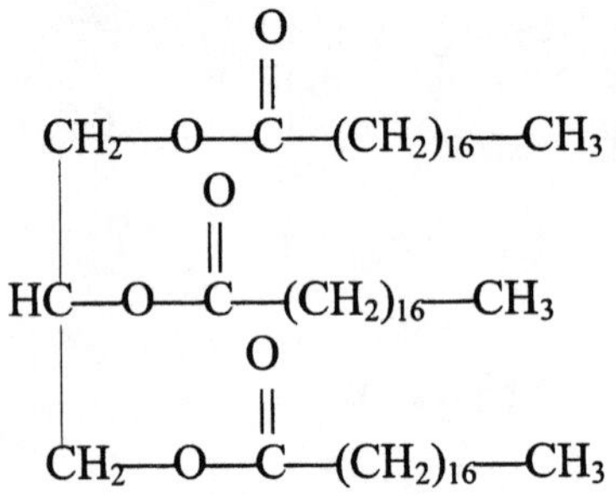

$$
\begin{array}{l}
CH_2-O-\overset{O}{\overset{\|}{C}}-(CH_2)_{16}-CH_3 \\
\;|\\
HC-O-\overset{O}{\overset{\|}{C}}-(CH_2)_{16}-CH_3 \\
\;|\\
CH_2-O-\overset{O}{\overset{\|}{C}}-(CH_2)_{16}-CH_3
\end{array}
$$

18.39 A triacylglycerol consists of glycerol and three fatty acids. A glycerophospholipid consists of glycerol, two fatty acids, a phosphate group, and an amino alcohol.

18.41

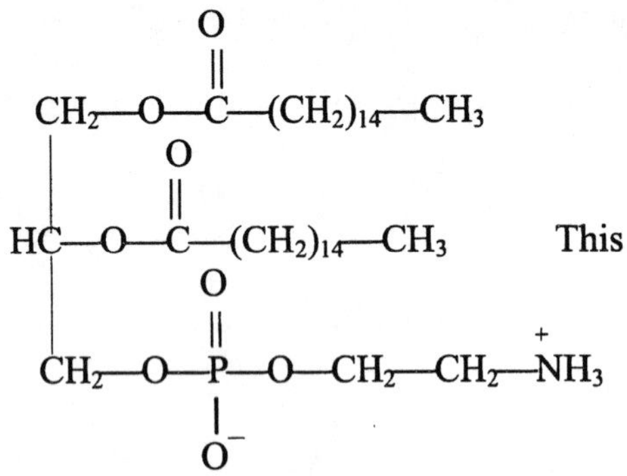

$$
\begin{array}{l}
CH_2-O-\overset{O}{\overset{\|}{C}}-(CH_2)_{14}-CH_3 \\
\;|\\
HC-O-\overset{O}{\overset{\|}{C}}-(CH_2)_{14}-CH_3 \\
\;|\\
CH_2-O-\overset{O}{\overset{\|}{P}}-O-CH_2-CH_2-\overset{+}{N}H_3 \\
\qquad\quad\;\; |\\
\qquad\quad\;\; O^-
\end{array}
$$

This is a cephalin.

18.43 This phospholipid is a cephalin. It contains glycerol, oleic acid, stearic acid, phosphate, and ethanolamine.

18.45 A glycerophospholipid consists of glycerol, two fatty acids, a phosphate group, and an amino alcohol. A sphingolipid contains the amino alcohol sphingosine instead of glycerol.

18.47

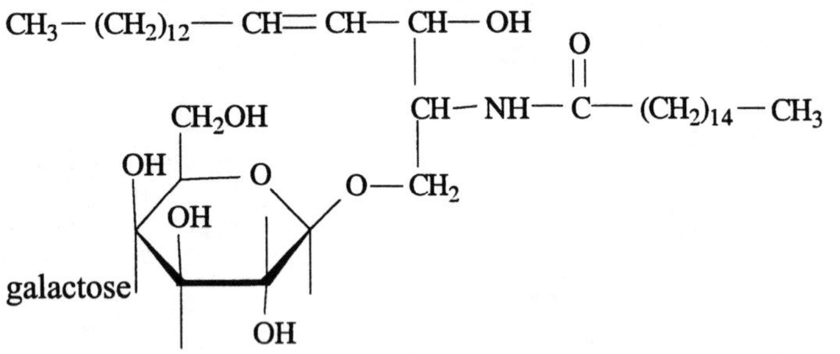

$$
\begin{array}{l}
CH_3-(CH_2)_{12}-CH{=}CH-CH-OH \\
\qquad\qquad\qquad\qquad\quad\;\; |\\
\qquad\qquad\qquad\qquad\quad\;\; CH-NH-\overset{O}{\overset{\|}{C}}-(CH_2)_{14}-CH_3 \\
\qquad\qquad\qquad\qquad\quad\;\; |\\
\qquad\qquad\qquad\quad\;\; O-CH_2
\end{array}
$$

galactose (CH_2OH, OH, O, OH, OH)

Palmitic acid

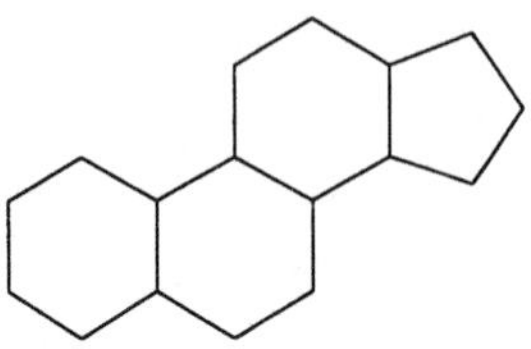

18.49

18.51 Bile salts emulsify fat globules, which makes the fat easier to digest by lipases.

18.53 Lipoproteins are large, spherically-shaped molecules that transport lipids in the bloodstream. They consist of an outer layer of phospholipids and proteins surrounding an inner core of hundreds of nonpolar lipids and cholesterol esters.

18.55 Chylomicrons have a lower density than VLDLs. They pick up triacylglycerols from the intestine, whereas VLDLs transport triacylglycerols synthesized in the liver.

18.57 "Bad" cholesterol is the cholesterol carried by LDLs to the tissues where it can form deposits called plaque, which can narrow the arteries.

18.59 Both estradiol and testosterone contain the steroid nucleus and a hydroxyl group. Testosterone has a ketone group, a double bond, and an extra methyl group. Estradiol has a benzene ring and a second hydroxyl group.

18.61 Testosterone is a male sex hormone.

18.63 Cell membranes contain a number of lipids: phospholipids that contain fatty acids with cis double bonds, glycolipids which are on the outside of the cell membrane and in animal membranes, cholesterol.

18.65 The function of the lipid bilayer in the plasma membrane is to keep the cell contents separated from the outside environment and to allow the cell to regulate the movement of substances into and out of the cell.

18.67 The peripheral proteins in the membrane emerge on the inner or outer surface only, whereas the integral proteins extend through the membrane to both surfaces.

18.69 The carbohydrates, as glycoproteins and glycolipids on the surface of cells, act as receptors for cell recognition and chemical messengers such as neurotransmitters.

18.71 Substances move through cell membrane by simple transport, facilitated transport, and active transport.

18.73 Beeswax and carnauba are waxes. Vegetable oil and capric triacylglycerol are triacylglycerols.

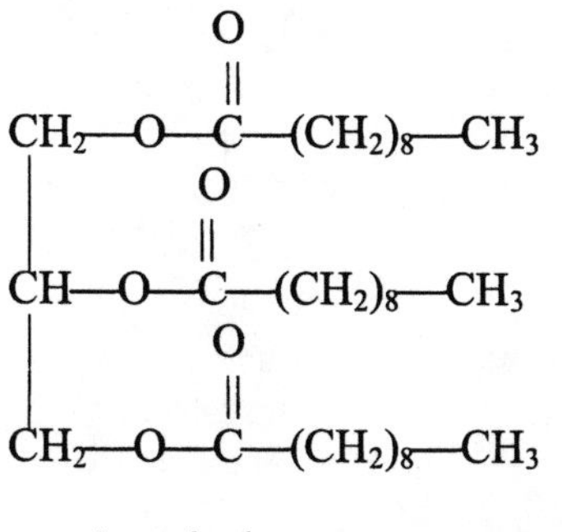

$$
\begin{array}{l}
CH_2-O-\overset{\overset{\displaystyle O}{\|}}{C}-(CH_2)_8-CH_3 \\
|\\
CH-O-\overset{\overset{\displaystyle O}{\|}}{C}-(CH_2)_8-CH_3 \\
|\\
CH_2-O-\overset{\overset{\displaystyle O}{\|}}{C}-(CH_2)_8-CH_3
\end{array}
$$

Capric triacylglycerol

18.75 **a.** A typical monounsaturated fatty acid has a cis double bond.

b. A trans fatty acid has a trans double bond with the alkyl groups on opposite sides of the double bond.

c.

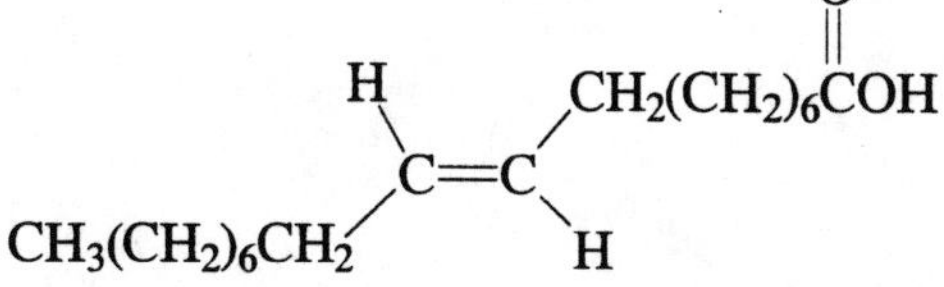

$$
\begin{array}{ccc}
H & & CH_2(CH_2)_6\overset{\overset{\displaystyle O}{\|}}{C}OH \\
 & C=C & \\
CH_3(CH_2)_6CH_2 & & H
\end{array}
$$

18.77

$$\begin{array}{l} \overset{\displaystyle O}{\overset{\|}{}} \\ CH_2-O-C-(CH_2)_{16}-CH_3 \\ | \qquad\quad \overset{\displaystyle O}{\|} \\ CH-O-C-(CH_2)_{16}-CH_3 \\ | \qquad\quad \overset{\displaystyle O}{\|} \\ CH_2-O-C-(CH_2)_{16}-CH_3 \end{array}$$

glyceryl tristearate

$$\begin{array}{l} \overset{\displaystyle O}{\overset{\|}{}} \\ CH_2-O-C-(CH_2)_{14}-CH_3 \\ | \qquad\quad \overset{\displaystyle O}{\|} \\ CH-O-C-(CH_2)_{14}-CH_3 \\ | \qquad\quad \overset{\displaystyle O}{\|} \qquad\qquad\qquad\qquad\quad CH_3 \\ | \qquad\qquad\qquad\qquad\qquad\qquad\quad |+ \\ CH_2-O-P-O-CH_2-CH_2-N-CH_3 \\ \qquad\quad\;\; | \qquad\qquad\qquad\qquad\quad\;\; | \\ \qquad\quad\; O^- \qquad\qquad\qquad\quad\;\; CH_3 \end{array}$$

lecithin

18.79 Stearic acid is a fatty acid. Sodium stearate is soap. Glyceryl tripalmitate, safflower oil, whale blubber and adipose tissue are triacylglycerols. Beeswax is a wax. Lecithin is a glycerophospholipid. Sphingomyelin is a sphingolipid. Cholesterol, progesterone, and cortisone are steroids.

18.81

a. 5 **b.** 1, 2, 3, 4 **c.** 2
d. 1, 2 **e.** 1, 2, 3, 4 **f.** 2, 3, 4, 6

18.83

a. 4 **b.** 3 **c.** 1
d. 4 **e.** 4 **f.** 3
g. 2 **h.** 1

19

Amines and Amides

Study Goals

- Name and write structural formulas of amines and amides.
- Describe the ionization of amines in water.
- Describe the boiling points of amines and amides compared to alkanes and alcohols.
- Describe the solubility of amines and amides in water.
- Write equations for the neutralization and amidation of amines.
- Describe acid and base hydrolysis of amides.

Think About It

1. Fish smell "fishy," but lemon juice removes the "fishy" odor. Why?

2. What functional groups are often found in tranquilizers and hallucinogens?

3. What is indicated by the codes such as 1-PETE on the bottom of plastic bottles and containers?

Key Terms

Match the key term with the correct statement show below.

a. heterocyclic amine **b.** amidation **c.** amine **d.** amide **e.** alkaloid

1. _____ A nitrogen-containing compound that is active physiologically

2. _____ A cyclic organic compound that contains one or more nitrogen atoms

3. _____ The reaction of a carboxylic acid and an amine

4. _____ The hydrolysis of this compound produces a carboxylic acid and an amine.

5. _____ An organic compound that contains an amino group

Answers **1.** e **2.** a **3.** b **4.** d **5.** c

19.1 Amines

- Amines are derivatives of ammonia (NH_3), in which alkyl or aromatic groups replace one or morehydrogen atoms.
- Amines are classified as primary, secondary, or tertiary when the nitrogen atom is bonded to one, two, or three alkyl or aromatic groups.

$CH_3—NH_2$	$CH_3—N(CH_3)—H$	$CH_3—N(CH_3)—CH_3$
Primary (1°)	Secondary (2°)	Tertiary (3°)

◆ **Learning Exercise 19.1**

Classify each of the following as a primary (1°), secondary (2°), or tertiary (3°) amine.

1. ____ $CH_3—NH—CH_2CH_3$ (H on N)

2. ____ cyclohexane ring with NH_2

3. ____ $CH_3—CH(NH_2)—C(=O)—OH$ ____

4. ____ $CH_3—CH_2—N(CH_3)—CH_3$

5. ____ $CH_3—CH_2—CH_2—CH_2—NH—CH_2—CH_3$

6. ____ benzene ring with NH_2 and CH_3 (meta)

Answers **1.** 2° **2.** 1° **3.** 1° **4.** 3° 5. **2°** 6. **1°**

19.2 Naming Amines

- Amines are usually named by common names in which the names of the alkyl group are listed alphabetically preceding the suffix *amine.*
- In the IUPAC system, the *e* of the alkane name of the main chain is replaced by *amine.* Alkyl groups attached to the N atom are named with the prefix *N*-.
- When another function group takes priority, —NH_2 is named as an amino substituent.
- The amine of benzene is named aniline.

	$CH_3—NH_2$	$CH_3—NH—CH_3$	$CH_3—N(CH_3)—CH_3$	benzene ring with NH_2
IUPAC:	methanamine	*N*-methylmethanamine	*N,N*,-dimethylmethanamine	aniline
Common:	methylamine	dimethylamine	trimethylamine	

- Many amines, which are prevalent in synthetic and naturally occurring compounds, have physiological activity.

◆ **Learning Exercise 19.2A**

Name each of the amines in problem 19.1.

1. ______________________ **2.** ______________________

3. ______________________ **4.** ______________________

5. ______________________ **6.** ______________________

Answers

1. ethylmethylamine; *N*-methylethanamine
2. cyclohexanamine
3. 2-aminopropanoic acid; *β*-aminopropionic acid
4. ethyldimethylamine; *N,N*-dimethylethanamine
5. butylethylamine; *N*-ethyl-1-butanamine
6. 3-methylaniline; *m*-methylaniline

◆ Learning Exercise 19.2B

Write the structural formulas of the following amines.

1. 2-propanamine
2. *N*-ethyl-*N*-methyl-1-aminobutane
3. 3-bromoaniline
4. *N*-methylaniline

Answers

1. $\mathrm{CH_3{-}CH(NH_2){-}CH_3}$ (NH₂ on the middle carbon)

$$\begin{array}{c} \mathrm{NH_2} \\ | \\ \mathrm{CH_3{-}CH{-}CH_3} \end{array}$$

2. $$\begin{array}{c} \qquad\qquad\mathrm{CH_3} \\ \qquad\qquad| \\ \mathrm{CH_3CH_2{-}N{-}CH_2CH_2CH_2CH_3} \end{array}$$

3. 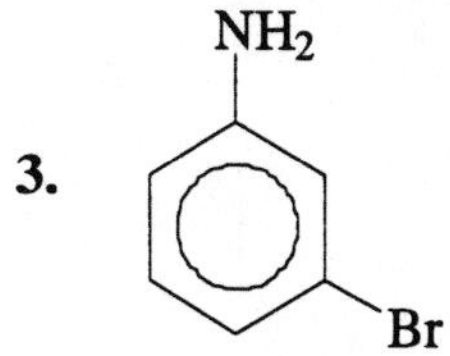

4. 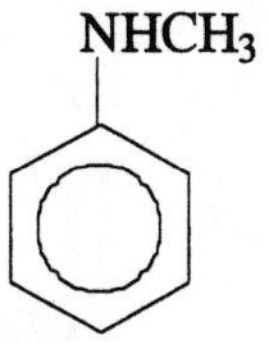 NHCH₃ (on benzene ring)

19.3 Physical Properties of Amines

- The N—H bonds in primary and secondary amines form hydrogen bonds.
- Amines have higher boiling points than hydrocarbons, but lower than alcohols of similar mass because the N atom is not as electronegative as the O atoms in alcohols.
- Hydrogen bonding allows amines with up to six carbon atoms to be soluble in water.

◆ Learning Exercise 19.3

Indicate the compound in each pair that has the higher boiling point.

1. $\mathrm{CH_3{-}NH_2}$ and $\mathrm{CH_3{-}OH}$ ________________
2. $\mathrm{CH_3{-}CH_2{-}CH_2}$ and $\mathrm{CH_3{-}CH_2{-}NH_2}$ ________________
3. $\mathrm{CH_3{-}CH_2{-}NH_2}$ and $\mathrm{CH_3{-}CH_2{-}CH_2{-}NH_2}$ ________________
4. $\begin{array}{c} \mathrm{H} \\ | \\ \mathrm{CH_3{-}N{-}CH_3} \end{array}$ and $\mathrm{CH_3{-}CH_2{-}NH_2}$ ________________

Answers

1. $\mathrm{CH_3{-}OH}$
2. $\mathrm{CH_3{-}CH_2{-}NH_2}$
3. $\mathrm{CH_3{-}CH_2{-}CH_2{-}NH_2}$
4. $\mathrm{CH_3{-}CH_2{-}NH_2}$

19.4 Amines React as Bases

- In water, amines act as weak bases by accepting protons from water to produce ammonium and hydroxide ions.

  $CH_3—NH_2 + H_2O \rightleftarrows CH_3—NH_3^+ + OH^-$
  methylamine *methylammonium hydroxide*

- Strong acids neutralize amines to yield ammonium salts.

  $CH_3—NH_2 + HCl \longrightarrow CH_3—NH_3^+ + Cl^-$
  methylamine *methylammonium chloride*

- When a carboxylic acid reacts with ammonia or an amine, an amide is produced.

$$CH_3—\overset{\overset{\displaystyle O}{\|}}{C}—OH + NH_3 \xrightarrow{\text{Heat}} CH_3—\overset{\overset{\displaystyle O}{\|}}{C}—NH_2 + H_2O$$

$$CH_3—\overset{\overset{\displaystyle O}{\|}}{C}—OH + NH_2—CH_3 \xrightarrow{\text{Heat}} CH_3—\overset{\overset{\displaystyle O}{\|}}{C}—NH—CH_3 + H_2O$$

◆ Learning Exercise 19.4A

Write the products of the following reactions.

1. $CH_3—CH_2—NH_2 + H_2O \rightleftarrows$
2. $CH_3—CH_2—CH_2—NH_2 + HCl \longrightarrow$
3. $CH_3CH_2—NH—CH_3 + HCl \longrightarrow$
4. (cyclohexane ring with NH_2) + HBr $\longrightarrow$
5. (cyclopentane ring with NH_2) + H_2O $\rightleftarrows$
6. $CH_3—CH_2—NH_3^+ Cl^- + NaOH \longrightarrow$

Answers

1. $CH_3—CH_2—NH_3^+ OH^-$
2. $CH_3—CH_2—CH_2—NH_3^+ Cl^-$
3. $CH_3—CH_2—\overset{+}{N}H_2—CH_3 Cl^-$
4. (cyclohexane ring with $NH_3^+ Br^-$)
5. (cyclopentane ring with $NH_3^+ OH^-$)
6. $CH_3—CH_2—NH_2 + NaCl + H_2O$

◆ Learning Exercise 19.4B

Write the structural formula(s) of the amides formed in each of the following reactions.

1. $CH_3-CH_2-\overset{\overset{O}{\|}}{C}-OH + NH_3 \xrightarrow{\text{Heat}}$

2. $C_6H_5-\overset{\overset{O}{\|}}{C}-OH + CH_3-NH_2 \xrightarrow{\text{Heat}}$

3. $CH_3-\overset{\overset{O}{\|}}{C}-OH + \overset{\overset{CH_3}{|}}{NH}-CH_3 \xrightarrow{\text{Heat}}$

Answers

1. $CH_3-CH_2-\overset{\overset{O}{\|}}{C}-NH_2$

2. $C_6H_5-\overset{\overset{O}{\|}}{C}-NH-CH_3$

3. $CH_3-\overset{\overset{O}{\|}}{C}-\overset{\overset{CH_3}{|}}{N}-CH_3$

19.5 Heterocyclic Amines and Alkaloids

- A heterocyclic amine is a cyclic compound containing one or more nitrogen atoms in the ring.
- Most heterocyclic amines contain 5 or 6 atoms.
- An alkaloid is a physiologically active amine obtained from plants.

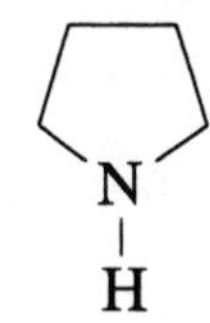

pyrrolidine

pyrrole

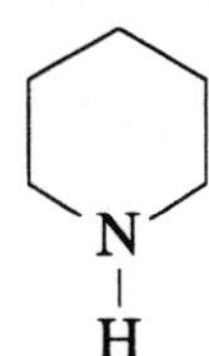

peperidine

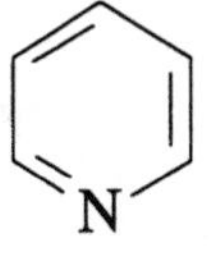

pyridine

◆ Learning Exercise 19.5

Match each of the following heterocyclic structures with the correct name.

a. 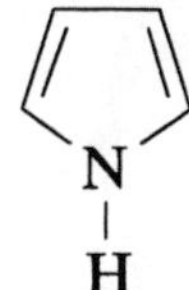

b. 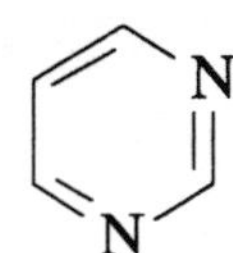

c. 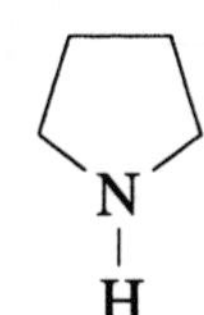

d. 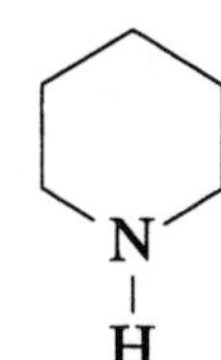

e. 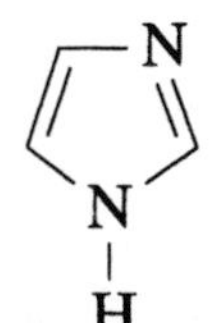

f. 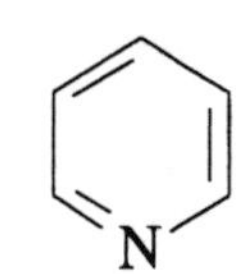

1. _____ pyrrolidine
2. _____ imidazole
3. _____ pyridine
4. _____ pyrrole
5. _____ pyrimidine
6. _____ piperidine

Answers **1.** c **2.** e **3.** f
4. a **5.** b **6.** d

19.6 Structures and Names of Amides

- Amides are derivatives of carboxylic acids in which an amine group replaces the —OH group in the acid.
- Amides are named by replacing the *ic acid* or *oic acid* ending by *amide.* When an alkyl group is attached to the *N* atom, it is listed as *N*-alkyl.

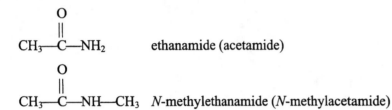

$$CH_3-\overset{\overset{\large O}{\|}}{C}-NH_2$$ ethanamide (acetamide)

$$CH_3-\overset{\overset{\large O}{\|}}{C}-NH-CH_3$$ *N*-methylethanamide (*N*-methylacetamide)

◆ Learning Exercise 19.6A

Name the following amides.

1. $CH_3-CH_2-\overset{\overset{\large O}{\|}}{C}-NH_2$ ____________________

2. 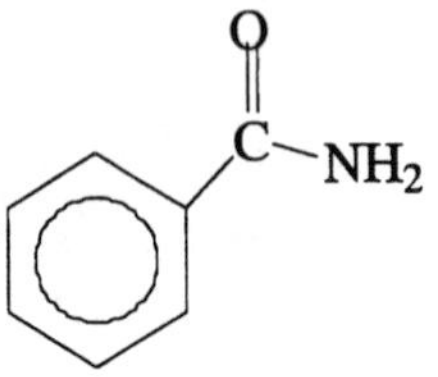 $C_6H_5-\overset{\overset{\large O}{\|}}{C}-NH_2$ (benzene ring) ____________________

3. $CH_3-CH_2-CH_2-CH_2-\overset{\overset{\large O}{\|}}{C}-NH-CH_3$ ____________________

4. $CH_3-\overset{\overset{\large O}{\|}}{C}-NH-CH_2-CH_3$ ____________________

5. 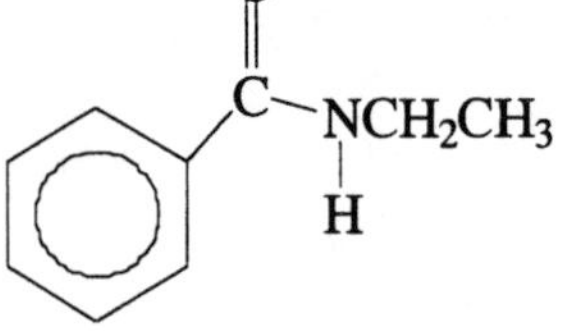 $C_6H_5-\overset{\overset{\large O}{\|}}{C}-\underset{\underset{\large H}{|}}{N}CH_2CH_3$ (benzene ring) ____________________

Answers
1. propanamide (propionamide)
2. benzamide
3. *N*-methylpentanamide
4. *N*-ethylethanamide (*N*-ethyl acetamide)
5. *N*-ethylbenzamide

◆ Learning Exercise 19.6B

Write the structural formulas for each of the following amides.

1. propanamide
2. *N*-methylbutanamide
3. *N*-methyl-3-chloropentanamide
4. benzamide

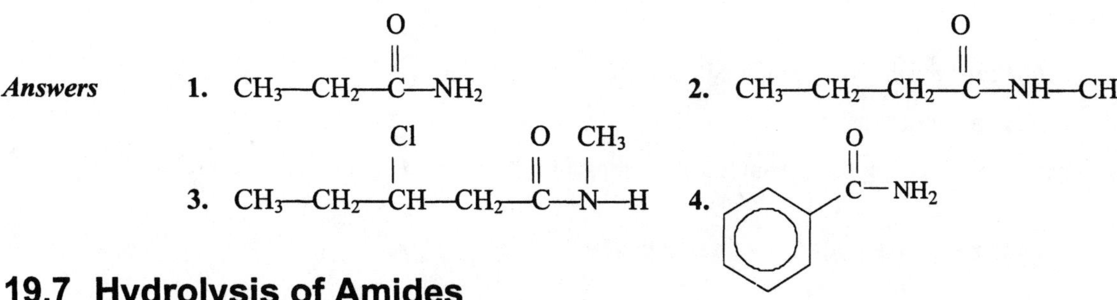

19.7 Hydrolysis of Amides

- Amides undergo acid and base hydrolysis to produce the carboxylic acid (or carboxylate salt) and the amine (or amine salt).

$$CH_3-\overset{\overset{\displaystyle O}{\|}}{C}-NH_2 + HCl + H_2O \longrightarrow CH_3-\overset{\overset{\displaystyle O}{\|}}{C}-OH + NH_4^+\, Cl^-$$

$$CH_3-\overset{\overset{\displaystyle O}{\|}}{C}-NH_2 + NaOH \longrightarrow CH_3-\overset{\overset{\displaystyle O}{\|}}{C}-O^-\, Na^+ + NH_3$$

◆ Learning Exercise 19.7

Write the structural formulas for the hydrolysis of each of the following with HCl and NaOH.

1. $CH_3-CH_2-\overset{\overset{\displaystyle O}{\|}}{C}-NH_2$

2. $CH_3-\overset{\overset{\displaystyle O}{\|}}{C}-NH-CH_2-CH_3$

Answers

1. (HCl) $CH_3-CH_2-\overset{\overset{\displaystyle O}{\|}}{C}-OH + NH_4^+\, Cl^-$ (NaOH)$CH_3-CH_2-\overset{\overset{\displaystyle O}{\|}}{C}-O^-\, Na^+ + NH_3$

2. (HCl) $CH_3-\overset{\overset{\displaystyle O}{\|}}{C}-OH + \overset{+}{N}H_3-CH_2-CH_3\, Cl^-$ (NaOH) $CH_3-\overset{\overset{\displaystyle O}{\|}}{C}-O^-\, Na^+ + NH_2-CH_2-CH_3$

Check List for Chapter 19

You are ready to take the practice test for Chapter 19. Be sure that you have accomplished the following learning goals for this chapter. If you are not sure, review the section listed at the end of the goal. Then apply your new skills and understanding to the practice test.

After studying Chapter 19, I can successfully:

____ Classify amines as primary, secondary, or tertiary (19.1).

____ Write the IUPAC and common names of amines and draw their condensed structural formulas (19.2).

____ Compare the boiling points and solubility of amines to alkanes and alcohols of similar mass (19.3).

____ Write equations for the ionization and neutralization of amines (19.4).

____ Identify heterocyclic amines (19.5).

____ Write the IUPAC and common names of amides and draw their condensed structural formulas (19.6).

____ Write equations for the hydrolysis of amines (19.7).

Practice Test for Chapter 19

Classify the amines in questions 1 through 6 as

A. primary amine **B.** secondary amine **C.** tertiary amine

1. ____ $CH_3-\overset{\overset{CH_3}{|}}{CH}-NH_2$

2. ____ $CH_3-CH_2-\overset{\overset{CH_3}{|}}{N}-CH_3$

3. ____ $CH_3-CH_2-\overset{\overset{NH_2}{|}}{CH}-CH_2-CH_3$

4. ____ $CH_3-\overset{\overset{H}{|}}{N}-CH_2-CH_3$

5. ____ $CH_3-\overset{\overset{CH_3}{|}}{CH}-CH_2-NH-\overset{\overset{CH_3}{|}}{CH}-CH_3$

6. ____ $CH_3-\underset{\underset{CH_3}{|}}{\overset{\overset{CH_3}{|}}{C}}-CH_2-NH_2$

Match the amines and amides in questions 7–11 with the following names.

A. ethyl dimethyl amine **B.** butanamide **C.** N-methylacetamide
D. benzamide **E.** N-ethylbutyramide

7. $CH_3-CH_2-\overset{\overset{CH_3}{|}}{N}-CH_3$

8. $CH_3-CH_2-CH_2-\overset{\overset{O}{\|}}{C}-NH_2$

9. $C_6H_5-\overset{\overset{O}{\|}}{C}-NH_2$

10. $CH_3—CH_2—CH_2—\overset{\overset{\displaystyle O}{\|}}{C}—NH—CH_2—CH_3$

11. $CH_3—\overset{\overset{\displaystyle O}{\|}}{C}—NH—CH_3$

In questions 12 through 15, identify the compound with the higher boiling point.

12. A. $CH_3—NH_2$ or **B.** $CH_3—OH$

13. A. $CH_3—NH—CH_3$ or **B.** $CH_3—CH_2—NH_2$

14. A. $CH_3—CH_2—CH_3$ or **B.** $CH_3—CH_2—NH_2$

15. A. $CH_3—CH_2—CH_2—OH$ or **B.** $CH_3—CH_2—CH_2—NH_2$

Match the products in questions 16 through 19 to the following reactions.

A. $CH_3—CH_2—\overset{\overset{\displaystyle O}{\|}}{C}—OH + NH_3$

B. $CH_3—CH_2—NH_3^+ \; Cl^-$

C. $CH_3—CH_2—CH_2—NH_3^+ \; OH^-$

D. $CH_3—\overset{\overset{\displaystyle O}{\|}}{C}—NH_2$

16. ____ ionization of 1-propanamine in water

17. ____ hydrolysis of propanamide

18. ____ reaction of ethanamine and hydrochloric acid

19. ____ amidation of acetic acid

20. Amines used in drugs are converted to their amine salt because the salt is

A. a solid at room temperature. **B.** soluble in water. **C.** odorless.
D. soluble in body fluids. **E.** all of these.

21. Heterocyclic amines are organic compounds that

A. have a ring of 5 or 6 atoms. **B.** contain one or more nitrogen atoms in a ring.
C. include pyrrolidine and pyrrole. **D.** include pyridine and pyrinmidine.
E. all of these

22. Alkaloids

A. are physiologically active nitrogen-containing compounds.
B. are produced by plants.
C. are used in anesthetics, antidepressants, and as stimulants.
D. are often habit forming.
E. all of these

Match the following alkaloids with their sources

A. caffeine **B.** nicotine **C.** morphine **E.** quinine

23. a painkiller from the Oriental poppy plant

24. obtained from the bark of the cinchona tree and used in the treatment of malaria

25. a stimulant obtained from the leaves of tobacco plants

26. a stimulant obtained from coffee beans and tea

Answers to the Practice Test

1. A	**2.** C	**3.** A	**4.** B	**5.** B
6. A	**7.** A	**8.** B	**9.** D	**10.** E
11. C	**12.** B	**13.** B	**14.** B	**15.** A
16. C.	**17.** A	**18.** B	**19.** D	**20.** E
21. E	**22.** E	**23.** C	**24.** E	**25.** B
26. A				

Answers and Solutions To Selected Text Problems

19.1 In a primary amine, there is one alkyl group (and two hydrogen atoms) attached to a nitrogen atom.

19.3 **a.** This is a primary (1°) amine; there is only one alkyl group attached to the nitrogen atom.
b. This is a secondary (2°) amine; there are two alkyl groups attached to the nitrogen atom.
c. This is a primary (1°) amine; there is only one alkyl group attached to the nitrogen atom.
d. This is a tertiary (3°) amine; there are three alkyl groups attached to the nitrogen atom.
e. This is a tertiary (3°) amine; there are three alkyl groups attached to the nitrogen atom.

19.5 The common name of an amine consists of naming the alkyl groups bonding to the nitrogen atom in alphabetical order. In the IUPAC name, the *e* in the alkane chain is replaced with *amine.*

a. An ethyl group attached to —NH_2 is ethylamine. In the IUPAC name, the *e* in ethane is replaced by *amine*: ethanamine.
b. Two alkyl groups attach to nitrogen as methyl and propyl for methylpropylamine. The IUPAC name based on the longer chain of propane with a methyl group attached to the nitrogen atoms is *N*-methyl-1-propanamine
c. Diethylmethylamine; *N*-methyl-*N*-ethylethanamine
d. Isopropylamine; 2-propanamine

19.7 The amine of benzene is called aniline. In amines where a more oxidized functional group takes priority, the —NH_2 group is named as an *amino* group and numbered.

a. 2-butanamine **b.** 2-chloroaniline
c. 3-aminopropanal **d.** *N*-ethylaniline

19.9 **a.** $CH_3—CH_2—NH_2$

b.

NHCH₃

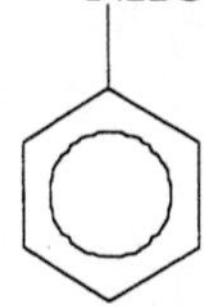

c.
$$\begin{array}{c} \qquad\qquad\qquad\qquad\qquad\qquad\qquad H \\ \qquad\qquad\qquad\qquad\qquad\qquad\qquad | \\ CH_3—CH_2—CH_2—CH_2—N—CH_2—CH_2—CH_3 \end{array}$$

d.

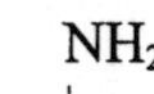

$$\begin{array}{l} \qquad\quad NH_2 \\ \qquad\quad | \\ CH_3—CH—CH_2—CH_2—CH_3 \end{array}$$

19.11 Amines have higher boiling points than hydrocarbons, but lower than alcohols of similar mass.

a. $CH_3—CH_2—OH$ **b.** $CH_3—CH_2—CH_2—NH_2$ **c.** $CH_3—CH_2—CH_2—NH_2$

19.13 Propylamine is a primary amine and can form two hydrogen bonds, which gives it the highest boiling point. Ethylmethylamine, a secondary amine, can form one hydrogen bond, and butane cannot form hydrogen bonds. Thus butane has the lowest boiling point of the three compounds.

19.15 Amines with one to five carbon atoms are soluble. The solubility in water of amines with longer carbon chains decreases.

a. yes **b.** yes **c.** no **d.** yes

19.17 Amines, which are weak bases, bond with a proton from water to give a hydroxide ion and an ammonium ion.

a. $CH_3—NH_2 + H_2O \rightleftarrows CH_3—NH_3^+ + OH^-$

b.
$$\begin{array}{l} \qquad\quad CH_3 \qquad\qquad\qquad\qquad CH_3 \\ \qquad\quad | \qquad\qquad\qquad\qquad\qquad | \\ CH_3—NH + H_2O \rightleftarrows CH_3—NH_2^+ + OH^- \end{array}$$

c. $C_6H_5—NH_2 + H_2O \rightleftarrows C_6H_5—NH_3^+ + OH^-$

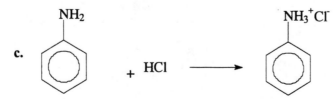

19.19 Amines, which are weak bases, combine with the proton from HCl to yield the ammonium chloride salt.

a. $CH_3—NH_2 + HCl \longrightarrow CH_3—NH_3^+\ Cl^-$

b.
$$\begin{array}{l} \qquad\quad CH_3 \qquad\qquad\qquad\qquad CH_3 \\ \qquad\quad | \qquad\qquad\qquad\qquad\qquad | \\ CH_3—NH + HCl \longrightarrow CH_3—NH_2^+\ Cl^- \end{array}$$

NH₂ NH₃⁺Cl⁻

c.

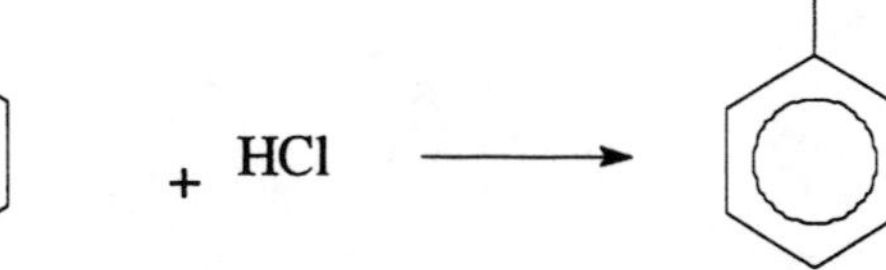

19.21 **a.** $H_2N-C_6H_4-\overset{O}{\overset{\|}{C}}-O-CH_2-CH_2-\overset{+}{N}(CH_2CH_3)_2-H \quad Cl^-$

b. Amine salts are soluble in body fluids.

19.23 **a.** Aniline is an amine.
b. An amine with three alkyl groups attached to the nitrogen atom
c. A nitrogen atom in a ring is a heterocyclic amine.
d. A nitrogen atom in a ring is a heterocyclic amine.

19.25 **c.** Pyrimidine has two nitrogen atoms in a ring of six atoms.
d. Pyrrole has one nitrogen atom in a ring of five atoms

19.27 The five-atom ring with one nitrogen atom and two double bonds is pyrrole.

19.29 Carboxylic acids react with amines to eliminate water and form amides.

a. $CH_3-\overset{O}{\overset{\|}{C}}-NH_2$

b. $CH_3-\overset{O}{\overset{\|}{C}}-NH-CH_2-CH_3$

c. $C_6H_5-\overset{O}{\overset{\|}{C}}-\overset{H}{\overset{|}{N}}-CH_2CH_2CH_3$

19.31 **a.** *N*-methylethanamide (*N*-methylacetamide). The *N*-methyl means that there is a one-carbon alkyl group attached to the nitrogen. Ethanamide tells us that the carbonyl portion has two carbon atoms.
b. Butanamide (butyramide) is a chain a four carbon atoms bonded to an amino group.
c. Methanamide (formamide)
d. *N*-methylbenzamide. The *N*-methyl means that there is a one-carbon alkyl group attached to the nitrogen. Benzamide tells us that this is the amid of benzoic acid.

19.33 **a.** This is an amide of propionic acid, which has three carbon atoms.

$CH_3-CH_2-\overset{O}{\overset{\|}{C}}-NH_2$

b. 2-methyl indicates that a methyl is bonded to carbon 2 in an amide chain of five carbon atoms.

$CH_3-CH_2-CH_2-\overset{CH_3}{\overset{|}{CH}}-\overset{O}{\overset{\|}{C}}-NH_2$

c. $H-\overset{\overset{O}{\|}}{C}-NH_2$

d. The nitrogen atom in *N, N*-ethylbenzamide is bonded to an ethyl group.

$C_6H_5-\overset{\overset{O}{\|}}{C}-\overset{\overset{H}{|}}{N}-CH_2CH_3$

e. The nitrogen atom is bonded to an ethyl group in *N*-ethylbutyramide.

$CH_3-CH_2-CH_2-\overset{\overset{O}{\|}}{C}-\overset{\overset{H}{|}}{N}-CH_2-CH_3$

19.35 **a.** Acetamide; primary amines have more hydrogen bonds and higher boiling points.
b. Propionamide can hydrogen bond, but butane cannot.
c. *N*-methylpropanamide can hydrogen bond, but *N,N*-dimethylpropanamide cannot.

19.37 Acid hydrolysis of amides gives the carboxylic acid and the amine salt.

a. $CH_3-COOH + NH_4^+\ Cl^-$ **b.** $CH_3-CH_2-COOH + NH_4^+\ Cl^-$

c. $CH_3-CH_2-CH_2-COOH + CH_3-NH_3^+\ Cl^-$ **d.** $C_6H_5-COOH\ +\ NH_4^+\ Cl^-$

e. $CH_3-CH_2-CH_2-CH_2-COOH + CH_3-CH_2-NH_3^+\ Cl^-$

19.39 $CH_3-CH_2-CH_2-NH_2$ Propanamine 1° $CH_3-CH_2-NH-CH_3$ *N*-methylmethanamine 2° $CH_3-\overset{\overset{CH_3}{|}}{N}-CH_3$ trimethylamine 3°

$CH_3-\overset{\overset{CH_3}{|}}{CH}-NH_2$ 2-propanamine 1°

19.41 **a.** $CH_3-CH_2-\overset{\overset{NH_2}{|}}{CH}-CH_2-CH_3$

b.

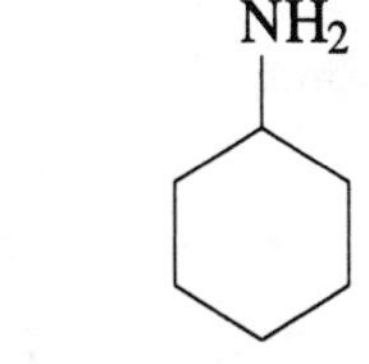

c. This is an ammonium salt with two methyl groups bonded to the nitrogen atom.

$CH_3-\overset{\overset{CH_3}{|}}{NH_2^+}\ Cl^-$

d. Three ethyl groups are bonded to a nitrogen atom

$CH_3-CH_2-\overset{\overset{CH_2-CH_3}{|}}{N}-CH_2-CH_3$

e. This six-carbon chain has a —NH_2 group on carbon 3 and an —OH on carbon 2.

$$\begin{array}{l} \quad\quad\;\; OH \;\; NH_2 \\ \quad\quad\;\;\; | \quad\;\; | \\ CH_3—CH—CH—CH_2—CH_2—CH_3 \end{array}$$

f.
$$\begin{array}{c} CH_3 \\ |^{+} \\ CH_3—N—CH_3\; Br^- \\ | \\ CH_3 \end{array}$$

g. Two methyl groups are bonded to the nitrogen of aniline.

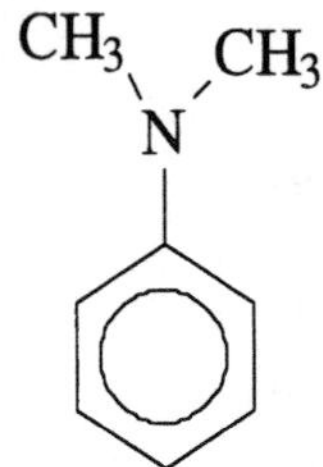

19.43 The smaller amines are more soluble in water.

a. ethylamine **b.** trimethylamine
c. butylamine **d.** $NH_2—CH_2—CH_2—CH_2—CH_2—CH_2—NH_2$

19.45 **a.** Quinine obtained from the bark of the cinchona tree is used in the treatment of malaria.
b. Nicotine is a stimulant found in cigarettes and cigars.
c. Caffeine is an alkaloid in coffee, tea, soft drinks, and chocolate.
d. Morphine and codeine are painkillers obtained from the oriental poppy plant.

19.47 **a.** An amine in water accepts a proton from water, which produces an ammonium ion and OH^-.

$CH_3—CH_2—NH_3^+\; OH^-$

b. The amine accepts a proton to give an ammonium salt: $CH_3—CH_2—NH_3^+\; Cl^-$

c. $CH_3—CH_2—\overset{+}{N}H_2—CH_3\; OH^-$

d. $CH_3—CH_2—\overset{+}{N}H_2—CH_3\; Cl^-$

e. An ammonium salt and a strong base produce the amine, a salt, and water.

$CH_3—CH_2—CH_2—NH_2\; + NaCl + H_2O$

f.
$$\begin{array}{l} \quad\quad\quad\quad\quad\quad\;\; CH_3 \\ \quad\quad\quad\quad\quad\quad\quad | \\ CH_3—CH_2—CH—NH_2\; + NaCl + H_2O \end{array}$$

19.49 carboxylic acid salt, aromatic, amine, haloaromatic

19.51 **a.** aromatic, amine, amide, carboxylic acid, cycloalkene
b. aromatic, ether, alcohol, amine
c. aromatic, carboxylic acid
d. phenol, amine, carboxylic acid
e. aromatic, ether, alcohol, amine, ketone
f. aromatic, amine

20

Amino Acids and Proteins

Study Goals

- Classify proteins by their functions in the cells.
- Draw the structures of amino acids.
- Draw the zwitterion forms of amino acids at the isoelectric point, and at pH levels above and below the isoelectric point.
- Write the structural formulas of di- and tripeptides.
- Identify the structural levels of proteins as primary, secondary, tertiary, and quaternary.
- Describe the effects of denaturation on the structure of proteins.

Think About It

1. What are some uses of protein in the body?

2. What are the units that make up a protein?

3. How do you obtain protein in your diet?

Key Terms

Match the following key terms with the correct statement shown below.

a. amino acid **b.** peptide bond **c.** denaturation
d. primary structure **e.** isoelectric point

1. ____ The order of amino acids in a protein

2. ____ The pH at which an amino acid has a net charge of zero

3. ____ The bond that connects amino acids in peptides and proteins

4. ____ The loss of secondary and tertiary protein structure caused by agents such as heat and acid

5. ____ The building block of proteins

Answers **1.** D **2.** E **3.** B **4.** C **5.** A

20.1 Functions of Proteins

- Some proteins are enzymes or hormones, while others are important in structure, transport, protection, storage, and contraction of muscles.

◆ Learning Exercise 20.1

Match one of the following functions of a protein with the examples below.

a. structural **b.** contractile **c.** storage **d.** transport
e. hormonal **f.** enzyme **g.** protection

1. _____ hemoglobin carries oxygen in blood
2. _____ amylase hydrolyzes starch
3. _____ egg albumin, a protein in egg white
4. _____ hormone that controls growth
5. _____ collagen makes up connective tissue
6. _____ immunoglobulin
7. _____ keratin, a major protein of hair
8. _____ lipoprotein carries lipids in blood

Answers	**1.** d	**2.** f	**3.** c	**4.** e
	5. a	**6.** g	**7.** a	**8.** d

20.2 Amino Acids

- A group of 20 amino acids provides the molecular building blocks of proteins.
- In an amino acid, a central (alpha) carbon is attached to an amino group, a carboxyl group, and a side chain or R group, which is a characteristic group for each amino acid.
- The particular R group makes each amino acid polar, nonpolar, acidic, or basic. Nonpolar amino acids contain hydrocarbon side chains, while polar amino acids contain electronegative atoms such as oxygen (—OH) or sulfur (—SH). Acidic side chains contain a carboxylic acid group and basic side chains contain an amino group ($—NH_2$).

◆ Learning Exercise 20.2

Using the appropriate R group, complete the structural formula of each of the following amino acids. Indicate whether the amino acid would be polar, nonpolar, acidic, or basic.

glycine (R = —H)

☐ (box above the central C); $H_3\overset{+}{N}—C—\overset{O}{\overset{\|}{C}}—OH$, with H below the central C

____

alanine (R = $—CH_3$)

☐ (box above the central C); $H_3\overset{+}{N}—C—\overset{O}{\overset{\|}{C}}—OH$, with H below the central C

____

serine (R = $-CH_2—OH$)

☐ (box above the central C); $H_3\overset{+}{N}—C—\overset{O}{\overset{\|}{C}}—OH$, with H below the central C

____

aspartic acid (R = $-CH_2—\overset{O}{\overset{\|}{C}}—OH$)

☐ (box above the central C); $H_3\overset{+}{N}—C—\overset{O}{\overset{\|}{C}}—OH$, with H below the central C

____

Answers

```
        H   O                              CH3 O
  +     |   ||                        +    |   ||
H3N—C—C—OH                        H3N—C—C—OH
        |                                  |
        H                                  H
nonpolar                          nonpolar

                                              O
                                              ||
    CH2—OH                         CH2—C—OH
    |   O                           |   O
  + |   ||                        + |   ||
H3N—C—C—OH                        H3N—C—C—OH
        |                                  |
        H                                  H
polar                             acidic
```

20.3 Amino Acids as Acids and Bases

- Amino acids exist as dipolar ions called zwitterions, which are neutral at the isoelectric point (pI).
- A zwitterion has a positive charge at pH levels below its pI and a negative charge at pH levels higher than its pI.

Study Note

Example: Glycine has an isoelectric point at a pH of 6.0. Write the zwitterion of glycine at its isoelectric point (pI), and at pH levels above and below its isoelectric point.

Solution: In more acidic solutions, glycine has a net positive charge, and in more basic solutions, a net negative charge.

$$H_3N^+—CH_2—COOH \xleftarrow{H^+} H_3N^+—CH_2—COO^- \xrightarrow{OH^-} H_2N—CH_2—COO^-$$

below pI *zwitterion of glycine* *above pI*

◆ Learning Exercise 20.3

Write the structure of the amino acids under the given conditions.

Zwitterion (pI)	H^+	OH^-
Alanine		
Serine		

Answers

Zwitterion (pI)	H^+	OH^-
Alanine $\overset{+}{H_3N}—\underset{}{\overset{CH_3}{\overset{\vert}{C}}}H—COO^-$	$\overset{+}{H_3N}—\overset{CH_3}{\overset{\vert}{C}}H—COOH$	$H_2N—\overset{CH_3}{\overset{\vert}{C}}H—COO^-$
Serine $\overset{+}{H_3N}—\overset{CH_2OH}{\overset{\vert}{C}}H—COO^-$	$\overset{+}{H_3N}—\overset{CH_2OH}{\overset{\vert}{C}}H—COOH$	$H_2N—\overset{CH_2OH}{\overset{\vert}{C}}H—COO^-$

20.4 Formation of Peptides

- A peptide bond is an amide bond between the carboxyl group of one amino acid and the amino group of the second.

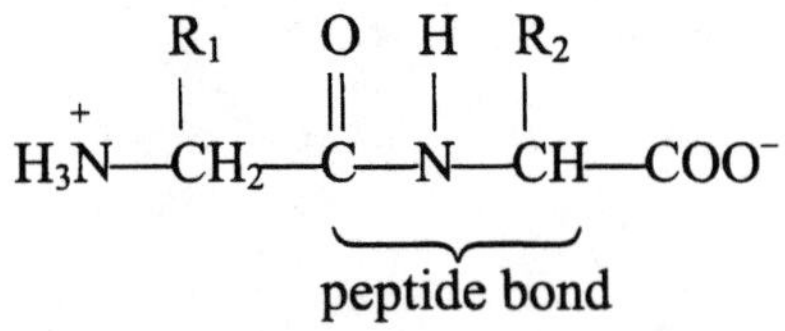

$$\overset{+}{H_3N}—\overset{R_1}{\overset{\vert}{CH_2}}—\underbrace{\overset{O}{\overset{\Vert}{C}}—\overset{H}{\overset{\vert}{N}}}_{\text{peptide bond}}—\overset{R_2}{\overset{\vert}{C}}H—COO^-$$

- Short chains of amino acids are called peptides. Long chains of amino acids are called proteins.

◆ Learning Exercise 20.4

Draw the structural formulas of the following di- and tripeptides.

1. serylglycine

2. cystylvaline

3. Gly-Ser-Cys

Answers

$$
\begin{array}{lllll}
 & \mathrm{HOCH_2} & \mathrm{O} & \mathrm{H} & \\
 & \;\;| & \| & | & \\
\textbf{1.}\;\; \overset{+}{\mathrm{H_2N}} - & \mathrm{CH} - & \mathrm{C} - & \mathrm{N} - & \mathrm{CH_2} - \mathrm{COO^-}
\end{array}
$$

$$
\begin{array}{llllll}
 & & & & \mathrm{CH_3} & \\
 & & & & | & \\
 & \mathrm{HSCH_2} & \mathrm{O} & \mathrm{H} & \mathrm{CH} - \mathrm{CH_3} & \\
 & \;\;| & \| & | & | & \\
\textbf{2.}\;\; \overset{+}{\mathrm{H_3N}} - & \mathrm{CH} - & \mathrm{C} - & \mathrm{N} - & \mathrm{CH} - & \mathrm{COO^-}
\end{array}
$$

$$
\begin{array}{llllllll}
 & & \mathrm{O} & & \mathrm{HOCH_2} & \mathrm{O} & & \mathrm{CH_2SH} \\
 & & \| & & | & \| & & | \\
\textbf{3.}\;\; \overset{+}{\mathrm{H_3N}} - & \mathrm{CH_2} - & \mathrm{C} - & \mathrm{N} - & \mathrm{CH} - & \mathrm{C} - & \mathrm{N} - & \mathrm{CH} - \mathrm{COO^-} \\
 & & & | & & & | & \\
 & & & \mathrm{H} & & & \mathrm{H} &
\end{array}
$$

20.5 Protein Structure: Primary and Secondary Levels

- The primary structure of a protein is the sequence of amino acids.
- In the secondary structure, hydrogen bonds between different sections of the peptide produce a characteristic shape such as an α-helix, β-pleated sheet, or a triple helix.
- Certain combinations of vegetables are complementary when the protein from one provides the missing amino acid in the other. For example, garbanzo beans and rice have complementary proteins because tryptophan—which is low in garbanzo beans—is provided by rice, and lysine—which is low in rice—is provided by garbanzo beans.

◆ Learning Exercise 20.5A

Identify the following descriptions of protein structure as primary or secondary structure.

1. _____ Hydrogen bonding forms an alpha (α)-helix.

2. _____ Hydrogen bonding occurs between C=O and N—H within a peptide chain.

3. _____ The order of amino acids, which are linked by peptide bonds

4. _____ Hydrogen bonds between protein chains form a pleated-sheet structure.

Answers **1.** secondary **2.** secondary **3.** primary **4.** secondary

◆ Learning Exercise 20.5B

Seeds, vegetables, and legumes are typically low in one or more of the essential amino acids: tryptophan, isoleucine, and lysine.

	tryptophan	isoleucine	lysine
sesame seeds	OK	LOW	LOW
sunflower seeds	OK	OK	LOW
garbanzo beans	LOW	OK	OK
rice	OK	OK	LOW
cornmeal	OK	OK	LOW

Indicate whether the following protein combinations are complementary or not.

1. _____ Sesame seeds and sunflower seeds
2. _____ Sunflower seeds and garbanzo beans
3. _____ Sunflower seeds, sesame seeds and garbanzo beans
4. _____ Sesame seeds and garbanzo beans
5. _____ Garbano beans and rice
6. _____ Cornmeal and garbanzo beans
7. _____ Rice and cornmeal

Answers
1. not complementary; both are low in lysine
2. complementary
3. complementary
4. complementary
5. complementary
6. complementary
7. not complementary; both are low in lysine

20.6 Protein Structure: Tertiary and Quaternary Levels

- In globular proteins, the polypeptide chain including its α-helical and β-pleated sheet regions folds upon itself to form a tertiary structure.
- In a tertiary structure, hydrophobic R groups are found on the inside and hydrophilic R groups on the outside surface. The tertiary structure is stabilized by interactions between R groups.
- In a quaternary structure, two or more subunits must combine for biological activity. They are held together by the same interactions found in tertiary structures.

◆ Learning Exercise 20.6

Identify the following descriptions of protein structure as tertiary or quaternary.

1. _____ A disulfide bond joining distant parts of a peptide
2. _____ The combination of four protein subunits
3. _____ Hydrophilic side groups seeking contact with water
4. _____ A salt bridge forms between two oppositely charged side chains.
5. _____ Hydrophobic side groups forming a nonpolar center

Answers 1. tertiary 2. quaternary 3. tertiary 4. tertiary 5. tertiary

20.7 Protein Hydrolysis and Denaturation

- Denaturation of a protein occurs when heat or other denaturing agents destroys the secondary and tertiary structure (but not the primary structure) of the protein until biological activity is lost.
- Denaturing agents include heat, acid, base, organic solvents, agitation, and metal ions.

◆ Learning Exercise 20.7

Indicate the denaturing agent in the following examples.

A. heat or UV light **B.** pH change **C.** organic solvent
D. heavy metal ions **E.** agitation

1. ____ Placing surgical instruments in a 120°C autoclave

2. ____ Whipping cream to make a desert topping

3. ____ Applying tannic acid to a burn

4. ____ Placing $AgNO_3$ drops in the eyes of newborns

5. ____ Using alcohol to disinfect a wound

6. ____ Using lactobacillus bacteria culture to produce acid that converts milk to yogurt.

Answers **1.** A **2.** E **3.** B **4.** D **5.** C **6.** B

Check List for Chapter 20

You are ready to take the practice test for Chapter 20 Be sure that you have accomplished the following learning goals for this chapter. If you are not sure, review the section listed at the end of the goal.

After studying Chapter 20, I can successfully:

____ Classify proteins by their functions in the cells (20.1).

____ Draw the structure for an amino acid (20.2).

____ Draw the zwitterion at the isoelectric point (pI), above, and below the pI (20.3).

____ Describe a peptide bond; draw the structure for a peptide (20.4).

____ Distinguish between the primary and secondary structures of a protein (20.5).

____ Distinguish between the tertiary and quaternary structures of a protein (20.6).

____ Describe the ways that denaturation affects the structure of a protein (20.7).

Practice Test for Chapter 20

1. Which amino acid is nonpolar?

A. serine **B.** aspartic acid **C.** valine **D.** cysteine **E.** glutamine

2. Which amino acid will form disulfide cross-links in a tertiary structure?

A. serine **B.** aspartic acid **C.** valine **D.** cysteine **E.** glutamine

3. Which amino acid has a basic side chain?

A. serine **B.** aspartic acid **C.** valine **D.** cysteine **E.** glutamine

4. All amino acids

A. have the same side chains.
B. form zwitterions.
C. have the same isoelectric points.
D. show hydrophobic tendencies.
E. are essential amino acids.

5. Essential amino acids

A. are the amino acids that must be supplied by the diet.
B. are not synthesized by the body.
C. are missing in incomplete proteins.
D. are present in proteins from animal sources.
E. all of the above

Use the following to answer questions for the amino acid alanine in questions 6 through 9.

A. $^{+}H_3N—CH(CH_3)—COO^-$ **B.** $H_2N—CH(CH_3)—COO^-$ **C.** $^{+}H_3N—CH(CH_3)—COOH$

6. _____ Alanine in its zwitterion form

7. _____ Alanine at a low pH

8. _____ Alanine at a high pH

9. _____ Alanine at its isoelectric point

10. The sequence Tyr-Ala-Gly
A. is a tripeptide. **B.** has two peptide bonds. **C.** has tyrosine with free —NH_2 end
D. has glycine with the free —COOH end **E.** all of these.

11. The type of bonding expected between lysine and aspartic acid is a
A. ionic bond. **B.** hydrogen bond. **C.** disulfide bond.
D. hydrophobic attraction. **E.** hydrophilic attraction.

12. What type of bond is used to form the α-helix structure of a protein?
A. peptide bond **B.** hydrogen bond **C.** ionic bond
D. disulfide bond **E.** hydrophobic attraction

13. What type of bonding places portions of the protein chain in the center of a tertiary structure?
A. peptide bonds **B.** ionic bonds **C.** disulfide bonds
D. hydrophobic attraction **E.** hydrophilic attraction

In questions 14 through 18, identify the protein structural levels that each of the following statements describe.

A. primary **B.** secondary **C.** tertiary
D. quaternary **E.** pentenary

14. ___ peptide bonds **15.** _____ a pleated sheet

16. ___ two or more protein subunits **17.** _____ an α-helix

18. ___ disulfide bonds

In questions 19 through 23, match the function of a protein with each example.

A. myoglobin in the muscles
B. α-keratin in skin
C. peptidase for protein hydrolysis
D. casein in milk

19. ___ enzyme

20. ___ structural

21. ___ transport

22. ___ storage

23. Denaturation of a protein
- **A.** occurs at a pH of 7.
- **B.** causes a change in protein structure.
- **C.** hydrolyzes a protein.
- **D.** oxidizes the protein.
- **E.** adds amino acids to a protein.

24. Which of the following will <u>not</u> cause denaturation?
A. 0°C **B.** $AgNO_3$ **C.** 80°C **D.** ethanol **E.** pH 1

Answers to the Practice Test

1. C	**2.** D	**3.** E	**4.** B	**5.** E
6. A	**7.** C	**8.** B	**9.** A	**10.** E
11. A	**12.** B	**13.** D	**14.** A	**15.** B
16. D	**17.** B	**18.** C	**19.** C	**20.** B
21. A	**22.** D	**23.** B	**24.** A	

Answers and Solutions to Selected Text Problems

20.1
- **a.** Hemoglobin, which carries oxygen in the blood, is a transport protein.
- **b.** Collagen, which is a major component of tendon and cartilage, is a structural protein.
- **c.** Keratin, which is found in hair, is a structural protein.
- **d.** Amylase, which catalyzes the breakdown of starch, is an enzyme.

20.3 All amino acids contain a carboxylic acid group and an amino group on the alpha carbon.

20.5

a.

CH₃ structure: $H_3\overset{+}{N}-CH(CH_3)(-CH_3)-C(=O)-O^-$

b.

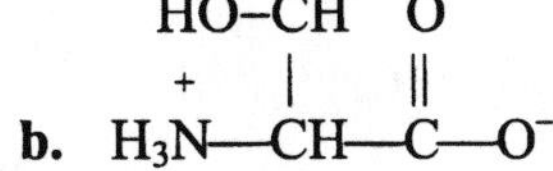

c.

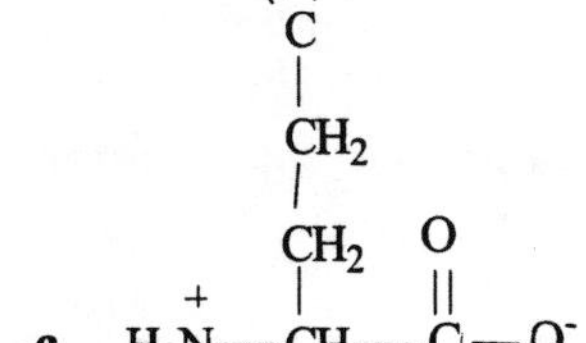

d.

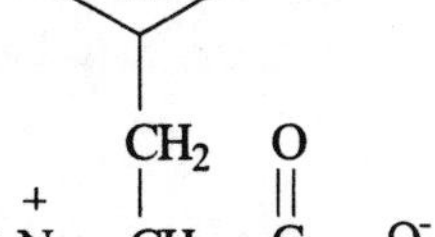

20.7
- **a.** Alanine, which has a methyl (hydrocarbon) side group, is nonpolar.
- **b.** Threonine has a side group that contains the polar —OH. Threonine is polar.
- **c.** Glutamic acid has a side group containing a polar carboxylic acid. Glutamic acid is acidic.
- **d.** Phenylalanine has a side group with a nonpolar benzene ring. Phenylalanine is nonpolar.

20.9 The abbreviations of most amino acids is derived from the first three letters in the name.

a. alanine **b.** valine **c.** lysine **d.** cysteine

20.11 In the L isomer, the —NH_2 is on the left side of the horizontal line of the Fischer projection; in D isomer, the —NH_2 group is on the right.

```
        COOH                        COOH
         |                           |
a. H2N ──┼── H          b.    H ─────┼───── NH2
         |                           |
         CH                         CH2SH
        /  \
     H3C    CH3
```

20.13 A zwitterion is formed when the H from the acid part of the amino acid is transferred to the amine portion of the amino acid. The resulting dipolar ion has an overall zero charge.

```
         H   O                           SH
  +      |   ||                          |
a. H3N──CH──CO-           b.             CH2  O
                                  +      |    ||  -
                                  H3N ── CH ── CO

          OH
          |
          CH2  O                         CH3  O
  +       |    ||  -             +       |    ||  -
c. H3N ── CH ── CO         d.   H3N ──── CH ── CO
```

20.15 At low pH (highly acidic), the $-COO^-$ of the zwitterion accepts a proton and the amino acid has a positive charge overall.

```
                                         SH
                                         |
         H   O                           CH2  O
  +      |   ||                  +       |    ||
a. H3N──CH──COH           b. H3N ──── CH ── COH

          OH
          |
          CH2  O                         CH3  O
  +       |    ||                +       |    ||
c. H3N ── CH ── COH        d.   H3N ──── CH ── COH
```

20.17
a. A negative charge means the zwitterion donated a proton from —NH_3^+, which occurs at pH levels above the isoelectric point
b. A positive charge means the zwitterion accepted a proton (H^+) from an acidic solution, which occurs at a pH level below the isoelectric point.
c. A zwitterion with a net charge of zero means that the pH level is equal to the isoelectric point.

20.19 In a peptide, the amino acids are joined by peptide bonds (amide bonds). The first amino acid has a free amine group, and the last one has a free carboxyl group.

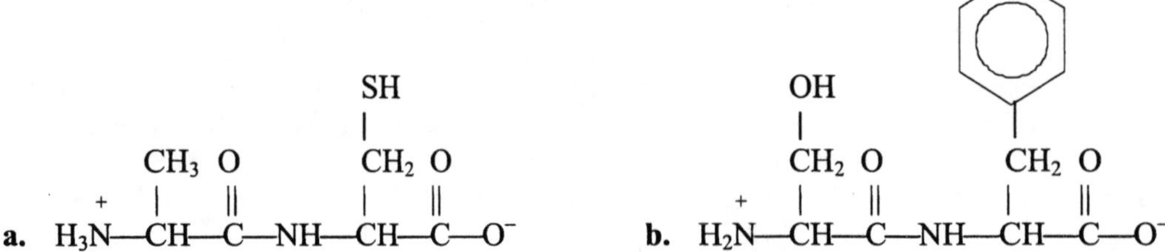

```
                                         CH3
                                          |
           O        CH3  O   H3C—CH   O
 +         ||        |   ||       |   ||
c. H3N—CH2—C—NH—CH—C—NH—CH—C—O⁻
```

```
                                        H
                                         \
                                          N
                   CH3                   (indole ring)
                    |
      CH3          CH2
       |            |
  H3C—CH   O  H3C—CH   O         CH2  O
   +   |   ||       |   ||          |   ||
d. H3N—CH—C—NH—CH—C—NH—CH—C—O⁻
```

20.21 The primary structure of a protein is the order of amino acids; the bonds that hold the amino acids together in a protein are amide or peptide bonds.

20.23 The possible primary structure of a tripeptide of one valine and two serines are:

Val-Ser-Ser, Ser-Val-Ser, and Ser-Ser-Val

20.25 When a protein forms a secondary structure, the amino acid chain arranges itself in space. The common secondary structures are: the alpha helix, the beta-pleated sheet, and the triple helix.

20.27 In an alpha helix, there are hydrogen bonds between the different turns of the helix, which preserves the helical shape of the protein. In a beta-pleated sheet, the hydrogen bonds occur between two protein chains that are side by side or between different parts of a long protein.

20.29
a. The two cysteine residues have —SH groups, which react to form a disulfide bond.
b. Glutamic acid is acidic, and lysine is basic; the two groups form an ionic bond, or salt bridge.
c. Serine has a polar —OH group that can form a hydrogen bond with the carboxyl group of aspartic acid.
d. Two leucine residues are hydrocarbon and nonpolar. They would have a hydrophobic interaction.

20.31
a. The R group of cysteine with the —SH group can form disulfide cross-links.
b. Leucine and valine are found on the inside of the protein since they have nonpolar side groups and are hydrophobic.
c. The cysteine and aspartic acid are on the outside of the protein since they are polar.
d. The order of the amino acid (the primary structure) provides the R groups, whose interactions determine the tertiary structure of the protein.

20.33 The complete hydrolysis of the tripeptide Gly-Ala-Ser will give the amino acids glycine (Gly) alanine (Ala) and serine (Ser).

20.35 Partial hydrolysis of the tetrapeptide His-Met-Gly-Val could give the following dipeptides:

Met-Gly; His-Met and Gly-Val

20.37 The primary level, the sequence of amino acids in the protein, is affected by hydrolysis.

20.39 **a.** Placing an egg in boiling water coagulates the protein of the egg by breaking the hydrogen bonds and disrupting the hydrophobic interactions.
b. Using an alcohol swab coagulates the protein of any bacteria present by forming new hydrogen bonds and disrupting hydrophobic interactions.
c. The heat from an autoclave will coagulate the protein of any bacteria on the surgical instruments by breaking the hydrogen bonds and disrupting the hydrophobic interactions.
d. Cauterization (heating) of a wound leads to coagulation of the protein and helps to close the wound by breaking hydrogen bonds and disrupting hydrophobic interactions.

20.41 **a.** A combination of rice and garbanzo beans contains all the three essential amino acids.
b. A combination of lima beans and cornmeal beans contains all the three essential amino acids.
c. A salad of garbanzo beans and lima beans does not contain all the three essential amino acids; it is deficient in tryptophan.
d. A combination of rice and lima beans contains all the three essential amino acids in lysine.
e. A combination of rice and oatmeal does not contain all the essential amino acids, it is deficient in lysine.
f. A combination of oatmeal and lima beans contains the three essential amino acids.

20.43 **a.** The secondary structure of a protein depends on hydrogen bonds to form a helix or a pleated sheet. The tertiary structure is determined by the interaction of R groups and determines the three dimensional structure of the protein.
b. Nonessential amino acids are synthesized by the body, but essential amino acids must be supplied by the diet.
c. Polar amino acids have hydrophilic side groups while nonpolar amino acids have hydrophobic side groups.
d. Dipeptides contain two amino acids, while tripeptides contain three.
e. An ionic bond is an interaction between a basic and acidic side group, a disulfide bond links the sulfides of two cysteines.
f. Fibrous proteins consist of three to seven alpha helixes coiled like a rope. Globular proteins form a compact spherical shape.
g. The alpha helix is the secondary shape like a spiral staircase or corkscrew. The beta-pleated sheet is a secondary structure that is formed by many proteins side by side, like a pleated sheet.
h. The tertiary structure of a protein is its three dimensional structure. In the quaternary structure, two or more peptide subunits are grouped.

20.45 **a.** β-keratins are fibrous proteins that provide structure to hair, wool, skin, and nails.
b. β-keratins have a high content of cysteine.

20.47 **a.**

$$H_3\overset{+}{N}-\underset{CH_2OH}{CH}-\overset{O}{\overset{\|}{C}}-\overset{H}{\overset{|}{N}}-\underset{(CH_2)_4NH_2}{CH}-\overset{O}{\overset{\|}{C}}-\overset{H}{\overset{|}{N}}-\underset{CH_2COOH}{CH}-\overset{O}{\overset{\|}{C}}-O^-$$

b. This segment contains polar R groups, which would be found on the surface of a globular proteins where they can hydrogen bond with water.

20.49 **a.** The β-pleated sheet is a secondary structure that contains high amount of Gly, Ala, and Ser, which have small side groups.
b. His, Met, and Leu are found predominantly in an α-helix secondary structure.

20.51 Serine is a polar amino acid, whereas valine is nonpolar. Valine would be in the center of the tertiary structure. However, serine would pull that part of the chain to the outside surface of the protein where valine forms hydrogen bonds with water.

20.53 **a.**
$$H_3\overset{+}{N}-\underset{\underset{CH_2OH}{|}}{CH}-\overset{\overset{O}{\|}}{C}-OH$$

b.
$$H_3\overset{+}{N}-\underset{\underset{CH_3}{|}}{CH}-\overset{\overset{O}{\|}}{C}-OH$$

c.
$$H_3\overset{+}{N}-\underset{\underset{\underset{\underset{NH_3}{|+}}{(CH_2)_4}}{|}}{CH}-\overset{\overset{O}{\|}}{C}-OH$$

21

Enzymes and Vitamins

Study Goals

- Classify enzymes according to the type of reaction they catalyze.
- Describe the lock-and-key and induced-fit models of enzyme action.
- Discuss the effect of changes in temperature, pH, and concentration of substrate on enzyme action.
- Describe the competitive, noncompetitive, and irreversible inhibition of enzymes.
- Discuss feedback control and regulation of enzyme action by allosteric enzymes.
- Identify the types of cofactors that are necessary for enzyme action.
- Describe the functions of vitamins and coenzymes.

Think About It

1. What are some functions of enzymes in the cells of the body?

2. Why are enzymes sensitive to high temperatures and low or high pH levels?

3. Why do we need vitamins?

Key Terms

Match the following key terms with the correct statement shown below.

a. lock-and-key theory **b.** vitamin **c.** inhibitor **d.** enzyme **e.** active site

1. _____ The portion of an enzyme structure where a substrate undergoes reaction

2. _____ A protein that catalyzes a biological reaction in the cells

3. _____ A model of enzyme action in which the substrate exactly fits the shape of an enzyme like a key fits into a lock

4. _____ A substance that makes an enzyme inactive by interfering with its ability to react with a substrate

5. _____ An organic compound essential for normal health and growth that must be obtained from the diet

Answers **1.** e **2.** d **3.** a **4.** c **5.** b

21.1 Biological Catalysts

- Enzymes are globular proteins that act as biological catalysts.
- Enzymes accelerate the rate of biological reactions by lowering the activation energy of a reaction.

◆ Learning Exercise 21.1A

Indicate whether each of the following characteristics of an enzyme is *true or false.*

An enzyme

1. ____ is a biological catalyst.

2. ____ functions at a low pH.

3. ____ usually does not change the equilibrium position of a reaction.

4. ____ is obtained from the diet.

5. ____ greatly increases the rate of a cellular reaction.

6. ____ is needed for every reaction that takes place in the cell.

7. ____ catalyzes at a faster rate at higher temperatures.

8. ____ functions best at mild conditions of pH 7.4 and 37°C.

9. ____ lowers the activation energy of a biological reaction.

10. ____ increases the rate of the forward reaction, but not the reverse.

Answers	**1.** T	**2.** F	**3.** T	**4.** F	**5.** T
	6. T	**7.** F	**8.** T	**9.** T	**10.** F

21.2 Names and Classification of Enzymes

- The names of most enzymes are indicated by their *ase* endings.
- Enzymes are classified by the type of reaction they catalyze: oxidoreductase, hydrolase, isomerase, transferase, lyase, or ligase.

◆ Learning Exercise 21.2A

Match the common name of each of the following enzymes with the description of the reaction.

1. dehydrogenase	**2.** oxidase	**3.** peptidase
4. decarboxylase	**5.** esterase	**6.** transaminase

a. ____ hydrolyzes the ester bonds in triacylglycerols to yield fatty acids and glycerol

b. ____ removes hydrogen from a substrate

c. ____ removes CO_2 from a substrate

d. ____ decomposes hydrogen peroxide to water and oxygen

e. ____ hydrolyzes peptide bonds during the digestion of proteins

f. ____ transfers an amino (NH_2) group from an amino acid to an α-keto acid

Answers	**a.** 5	**b.** 1	**c.** 4	**d.** 2	**e.** 3	**f.** 6

◆ Learning Exercise 21.2B

Match the IUPAC classification for enzymes with each of the following types of reactions.

1 oxidoreductase	2. transferase	3. hydrolase
4. lyase	5. isomerase	6. ligase

a. _____ combines small molecules using energy from ATP

b. _____ transfers phosphate groups

c. _____ hydrolyzes a disaccharide into two glucose units

d. _____ converts a substrate to an isomer of the substrate.

e. _____ adds hydrogen to a substrate

f. _____ removes H_2O from a substrate.

g. _____ adds oxygen to a substrate

h. _____ converts a cis structure to a trans structure

Answers **a.** 6 **b.** 2 **c.** 3 **d.** 5 **e.** 1 **f.** 4 **g.** 1 **h.** 5

21.3 Enzymes as Catalysts

- Within the structure of the enzyme, there is a small pocket called the active site, which has a specific shape that fits a specific substrate.
- In the lock-and-key model or the induced-fit model, an enzyme and substrate form an enzyme-substrate complex so the reaction of the substrate can be catalyzed at the active site.

◆ Learning Exercise 21.3A

Match the terms (A) active site, (B) substrate, (C) enzyme-substrate complex, (D) lock-and-key, and (E) induced-fit with the following descriptions:

1. _____ the combination of an enzyme with a substrate
2. _____ a model of enzyme action in which the rigid shape of the active site exactly fits the shape of the substrate
3. _____ has a tertiary structure that fits the structure of the active site
4. _____ a model of enzyme action in which the shape of the active site adjusts to fit the shape of a substrate
5. _____ the portion of an enzyme that binds to the substrate and catalyzes the reaction

Answers **1.** C **2.** D **3.** B **4.** E **5.** A

◆ Learning Exercise 21.3B

Write an equation to illustrate the following:

1. The formation of an enzyme-substrate complex. ______________________
2. The conversion of enzyme-substrate complex to product. ______________________

Answers **1.** $E + S \rightleftarrows ES$ **2.** $ES \longrightarrow E + P$

21.4 Factors Affecting Enzyme Activity

- Enzymes are most effective at optimum temperature and pH. The rate of an enzyme reaction decreases considerably at temperatures and pH above or below the optimum.
- An enzyme can be made inactive by changes in pH, temperature, or by chemical compounds called inhibitors.
- An increase in substrate concentration increases the reaction rate of an enzyme-catalyzed reaction until all the enzyme molecules combine with substrate.

◆ Learning Exercise 21.4A

Urease, which has an optimum pH of 5, catalyzes the hydrolysis of urea to ammonia and CO_2 in the liver.

$$H_2N-\overset{\overset{\displaystyle O}{\|}}{C}-NH_2 + H_2O \xrightarrow{\text{urease}} 2NH_3 + CO_2$$

Draw a graph to represent the effects of each of the following on enzyme activity. Indicate the optimum pH and optimum temperature.

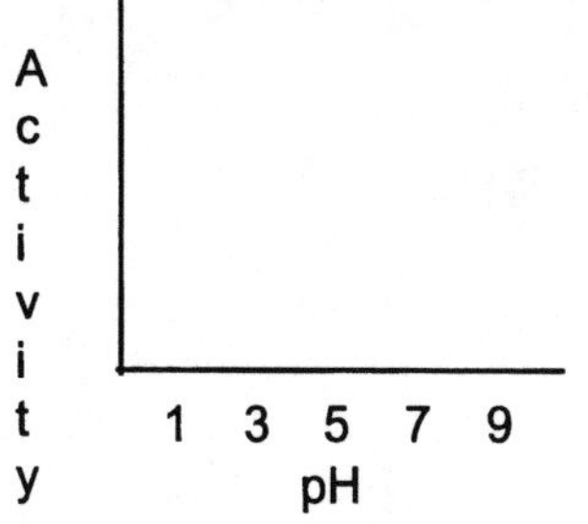

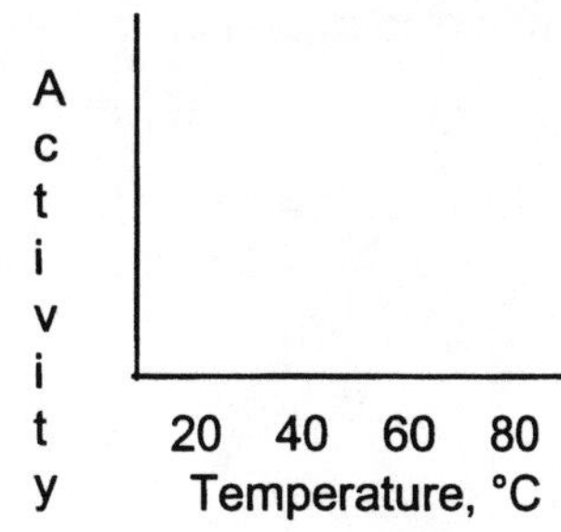

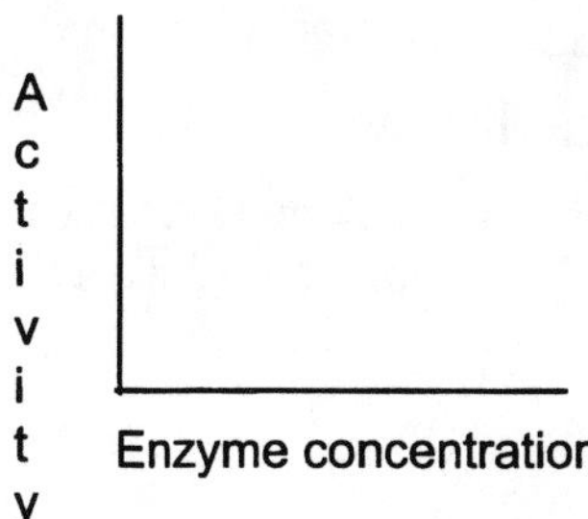

How is the rate of the urease-catalyzed reaction affected by each of the following?

a. increases **b.** decreases **c.** not changed

1. ____ adding more urea when an excess of enzyme is present

2. ____ running the reaction at pH 8

3. ____ lowering the temperature to 0°C

4. ____ running the reaction at 85°C

5. ____ increasing the concentration of urease for a specific amount of urea

6. ____ adjusting pH to the optimum

Answers

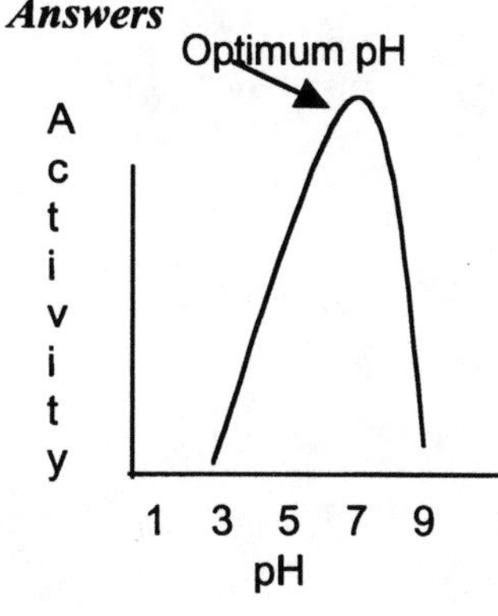

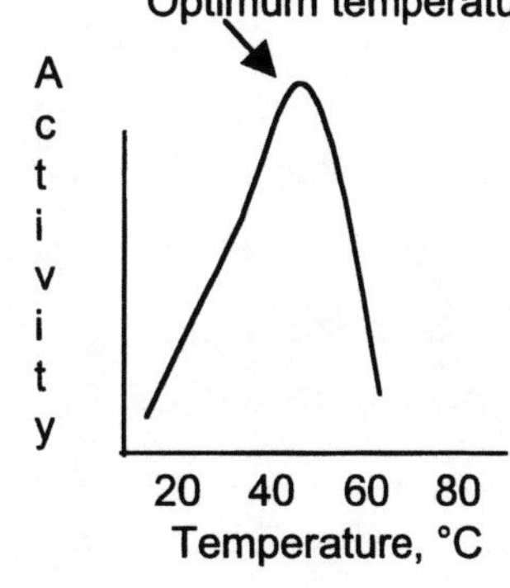

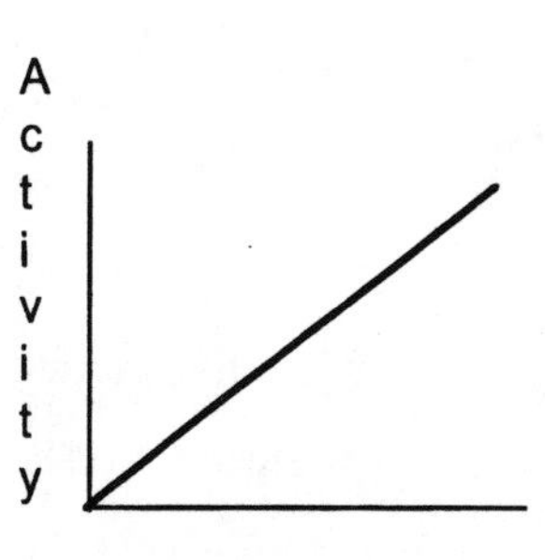

1. a **2.** b **3.** b **4.** b **5.** a **6.** a

21.5 Enzyme Inhibition

- A competitive inhibitor has a structure similar to the substrate and competes for the active site. When the active site is occupied by a competitive inhibitor, the enzyme cannot catalyze the reaction of the substrate.
- A noncompetitive inhibitor attaches elsewhere on the enzyme changing the shape of both the enzyme and the active site. As long as the noncompetitive inhibitor is attached to the enzyme, the altered active site cannot bind with substrate.

◆ Learning Exercise 21.5

Identify each of the following as characteristic of competitive inhibition (C), noncompetitive inhibition (N), or irreversible inhibition (I).

1. _____ An inhibitor binds to the surface of the enzyme away from the active site.
2. _____ An inhibitor resembling the substrate molecule blocks the active site on the enzyme.
3. _____ An inhibition causes permanent damage to the enzyme with a total loss of biological activity.
4. _____ The action of this inhibitor can be reversed by adding more substrate.
5. _____ Increasing substrate concentration does not change the effect of this inhibition.
6. _____ Sulfanilamide stops bacterial infections because its structure is similar to PABA (p-aminobenzoic acid), which is essential for bacterial growth.

Answers **1.** N **2.** C **3.** I **4.** C **5.** N **6.** C

21.6 Regulation of Enzyme Activity

- Many digestive enzymes are produced and stored as inactive forms called zymogens, which are activated at a later time.
- Hormones such as insulin and enzymes that catalyze blood clotting are synthesized as zymogens.
- When allosteric enzymes bind regulator molecules on a different part of the enzyme, there is a change in the shape of the enzyme and the active site. A positive regulator speeds up a reaction, and a negative regulator slows down a reaction.
- In feedback control, the end product of an enzyme-catalyzed sequence acts as a negative regulator and binds to the first enzyme in the sequence, which slows the rate of catalytic activity.

◆ Learning Exercise 21.6

Match the following characteristics of types of enzyme regulation.

(Z) zymogen (A) allosteric enzyme (P) positive regulator
(N) negative regulator (F) feedback control

1. _____ An enzyme that binds molecules at a site that is not the active site to speed up or slow down the rate of enzyme activity
2. _____ The end product of a reaction sequence binds to the first enzyme in the pathway.
3. _____ A molecule that slows down reaction by preventing proper binding to the substrate
4. _____ An inactive form of an enzyme that is activated by removing a peptide section
5. _____ A molecule that binds at a site different from the active site to speed up the reaction

Answers **1.** A **2.** F **3.** N **4.** Z **5.** P

21.7 Enzyme Cofactors

- Simple enzymes are biologically active as a protein only, whereas other enzymes require a cofactor.
- A cofactor may be a metal ion such as Cu^{2+} or Fe^{2+}, or an organic compound called a coenzyme, usually a vitamin.

◆ Learning Exercise 21.7

Indicate whether each statement describes a simple enzyme or a protein that requires a cofactor.

1. ___________ An enzyme consisting only of protein

2. ___________ An enzyme requiring magnesium ion for activity

3 ___________ An enzyme containing a sugar group

4. ___________ An enzyme that gives only amino acids upon hydrolysis

5. ___________ An enzyme that requires zinc ions for activity

Answers **1.** simple **2.** requires a cofactor **3.** requires a cofactor
4. simple **5.** requires a cofactor

21.8 Vitamins and Coenzymes

- Vitamins are organic molecules that are essential for proper health.
- Vitamins must be obtained from the diet because they are not synthesized in the body.
- Vitamins B and C are classified as water-soluble; vitamins A, D, E, and K are fat-soluble vitamins.
- Many water-soluble vitamins function as coenzymes.

◆ Learning Exercise 21.7A

Identify the water-soluble vitamin associated with each of the following:

a. thiamin (B_1) **b.** riboflavin (B_2) **c.** niacin (B_3)
d. cobalamin (B_{12}) **e.** ascorbic acid (C) **f.** pantothenic acid (B_5)

1. ____ collagen formation **2.** ____ coenzyme for NAD^+

3. ____ pellagra **4.** ____ part of coenzyme A

5. ____ FAD and FMN **6.** ____ scurvy

Answers **1.** e **2.** c **3.** c **4.** f **5.** b **6.** e

◆ Learning Exercise 21.7B

Identify the fat-soluble vitamin associated with each of the following:

a. vitamin A **b.** vitamin D **c.** vitamin E **d.** vitamin K

1. ____ prevents oxidation of fatty acids **2.** ____ blood clotting

3. ____ night vision **4.** ____ rickets

5. ____ formed in skin from sunlight **6.** ____ derived from cholesterol

Answers **1.** c **2.** d **3.** a **4.** b **5.** b **6.** b

Check List for Chapter 21

You are ready to take the practice test for Chapter 21. Be sure that you have accomplished the following learning goals for this chapter. If you are not sure, review the section listed at the end of the goal.

After studying Chapter 21, I can successfully:

____ Classify enzymes according to the type of reaction they catalyze (21.2).

____ Describe the lock-and-key and induced fit models of enzyme action (21.3).

____ Discuss the effect of changes in temperature, pH, and concentration of substrate on enzyme action (21.4).

____ Describe the reversible and irreversible inhibition of enzymes (21.5).

____ Discuss feedback control and regulation of enzyme action (21.6).

____ Identify the types of cofactors that are necessary for enzyme action (21.7).

____ Describe the functions of vitamins as coenzymes (21.8).

Practice Test for Chapter 21

1. Enzymes are ____.

   A. biological catalysts. **B.** polysaccharides. **C.** insoluble in water.
   D. always contain a cofactor. **E.** named with an "ose" ending.

Classify the enzymes described in questions 2 through 5 as (A) simple or (B) require a cofactor

2. ____ An enzyme that yields amino acids and a glucose molecule on analysis

3. ____ An enzyme consisting of protein only

4. ____ An enzyme requiring zinc ion for activation

5. ____ An enzyme containing vitamin K

For problems 6 through 10, select answers from the following: (**E** = enzyme; **S** = substrate; **P** = product)

A. $S \longrightarrow P$ **B.** $EP \longrightarrow E + P$ **C.** $E + S \longrightarrow ES$
D. $ES \longrightarrow EP$ **E.** $EP \longrightarrow ES$

6. ____ The enzymatic reaction occurring at the active site

7. ____ The release of product from the enzyme

8. ____ The first step in the lock-and-key theory of enzyme action

9. ____ The formation of the enzyme-substrate complex

10. ____ The final step in the lock-and-key theory of enzyme action

In problems 11 through 15, match the names of enzymes with a reaction they each catalyze.

A. decarboxylase **B.** isomerase **C.** dehydrogenase
D. lipase **E.** sucrase

11. ____ $CH_3—CH(OH)—COOH \longrightarrow CH_3—C(=O)—COOH$

12. ____ sucrose + $H_2O \longrightarrow$ glucose and fructose

13. ____ $CH_3\text{—}\overset{\overset{\large O}{\|}}{C}\text{—}COOH \longrightarrow CH_3COOH + CO_2$

14. ____ fructose ⟶ glucose

15. ____ triglyceride + $3H_2O$ ⟶ fatty acids and glycerol

For problems 16 through 20 select your answers from

A. Increases the rate of reaction
B. Decreases the rate of reaction
C. Denatures the enzyme and no reaction occurs

16. ____ Setting the reaction tube in a beaker of water at 100°C

17. ____ Adding substrate to the reaction vessel

18. ____ Running the reaction at 10°C

19. ____ Adding ethanol to the reaction system

20. ____ Adjusting the pH to optimum pH

For problems 21 through 25, identify each description of inhibition as

A. competitive
B. noncompetitive

21. ____ An alteration in the conformation of the enzyme

22. ____ A molecule closely resembling the substrate interferes with activity

23. ____ The inhibition can be reversed by increasing substrate concentration.

24. ____ The heavy metal ion, Pb^{2+}, bonds with an —SH side group.

25. ____ The inhibition is not affected by increased substrate concentration.

26. ____ The presence of zinc in the enzyme called alcohol dehydrogenase classifies a protein as

A. simple
B. requiring a cofactor
C. hormonal
D. structural
E. secondary

Answers to the Practice Test

1. A	2. B	3. A	4. B	5. B
6. D	7. B	8. C	9. C	10. B
11. C	12. E	13. A	14. B	15. D
16. C	17. A	18. B	19. C	20. A
21. B	22. A	23. A	24. B	25. B
26. B				

Answers and Solutions to Selected Text Problems

21.1 The chemical reactions can occur without enzymes, but the rates are too slow. Catalyzed reactions, which are many times faster, provide the amounts of products needed by the cell at a particular time.

21.3 **a.** Oxidoreductases catalyze oxidation and reduction.
b. Transferases move groups such as amino groups or phosphate groups from one substance to another.
c. Hydrolases use water to split bonds in molecules such as carbohydrates, peptides, and lipids.

21.5 **a.** A hydrolase enzyme would catalyze the hydrolysis of sucrose.
b. An oxidoreductase enzyme would catalyze the addition of oxygen (oxidation).
c. An isomerase enzyme would catalyze converting glucose to fructose.
d. A transferase enzyme would catalyze moving an amino group.

21.7 **a.** A lyase such as a decarboxylase removes CO_2 from a molecule.
b. The transfer of an amino group to another molecule would be catalyzed by a transferase.

21.9 **a.** Succinate oxidase catalyzes the oxidation of succinate.
b. Fumarate hydrase catalyzes the addition of water to fumarate.
c. Alcohol dehydrogenase removes 2H from an alcohol.

21.11 **a.** An enzyme has a tertiary structure that recognized the substrate.
b. The combination of the enzyme and substrate is the enzyme-substrate complex.
c. The substrate has a structure that complements the structure of the enzyme.

21.13 **a.** The equation for an enzyme-catalyzed reaction is:

$$E + S \rightleftarrows ES \longrightarrow E + P$$

E = enzyme, S = substrate, ES = enzyme-substrate complex, P = products

b. The active site is a region or pocket within the tertiary structure of an enzyme that accepts the substrate, aligns the substrate for reaction, and catalyzes the reaction.

21.15 Isoenzymes are slightly different forms of an enzyme that catalyze the same reaction in different organs and tissues of the body.

21.17 A doctor might run tests for the enzymes CK, LDH, and AST to determine if the patient had a heart attack.

21.19 **a.** Decreasing the substrate concentration decreases the rate of reaction.
b. Running the reaction at a pH below optimum pH will decrease the rate of reaction.
c. Temperature above 37°C (optimum pH) will denature the enzymes and decrease the rate of reaction.
d. Increasing the enzyme concentration would increase the rate of reaction.

21.21 pepsin, pH 2; urease, pH 5; trypsin , pH 8

21.23 **a.** If the inhibitor has a structure similar to the structure of the substrate, the inhibitor is competitive
b. If adding more substrate cannot reverse the effect of the inhibitor, the inhibitor is noncompetitive.
c. If the inhibitor competes with the substrate for the active site, it is a competitive inhibitor.
d. If the structure of the inhibitor is not similar to the structure of the substrate, the inhibitor is noncompetitive.
e. If adding more substrate reverses inhibition, the inhibitor is competitive.

21.25 **a.** Methanol has the structural formula CH_3—OH whereas ethanol is CH_3—CH_2—OH.
b. Ethanol has a structure similar to methanol and could compete for the active site.
c. Ethanol is a competitive inhibitor of methanol oxidation.

21.27 Digestive enzymes are proteases and would digest the proteins of the organ where they are produced if they were active immediately upon synthesis.

21.29 In feedback inhibition, the product binds to the first enzyme in a series changing the shape of the active site. If the active site can no longer bind the substrate effectively, the reaction stops.

21.31 When a regulator molecule binds to an allosteric site, the shape of the enzyme is altered, which makes the active site more or less reactive and thereby increases or decreases the rate of the reaction.

21.33 **a.** 3 A negative regulator binds to the allosteric site and slows down the reaction.
b. 4 Typically the first enzyme in a reaction sequence is an allosteric enzyme, which regulates the flow of substrates through the sequence to yield end product.
c. 1 A zymogen is an inactive form of an enzyme.

21.35 **a.** The active form of this enzyme requires a cofactor.
b. The active form of this enzyme requires a cofactor.
c. A simple enzyme is active as a protein.

21.37 **a.** THF **b.** NAD^+

21.39 **a.** Pantothenic acid (vitamin B_5) is part of coenzyme A.
b. Tetrahydrofolate (THF) is a reduced form of folic acid.
c. Niacin (vitamin B_3) is a component of NAD^+.

21.41 **a.** A deficiency of vitamin D or cholecalciferol can lead to rickets.
b. A deficiency of ascorbic acid or vitamin C can lead to scurvy.
c. A deficiency of niacin or vitamin B_3 can lead to pellagra.

21.43 Vitamin B_6 is a water-soluble vitamin, which means that each day any excess of vitamin B_6 is eliminated from the body.

21.45 The side chain —CH_2OH on the ring is oxidized to —CHO, and the other —CH_2OH forms a phosphate ester.

21.47 The many different reactions that take place in cells require different enzymes because enzymes react with only a certain type of substrate.

21.49 When exposed to conditions of strong acids or bases, or high temperatures, enzymatic proteins are denatured rapidly, causing a loss of tertiary structure and catalytic activity.

21.51 **a.** The reactant is lactose and the products are glucose and galactose.

b. \

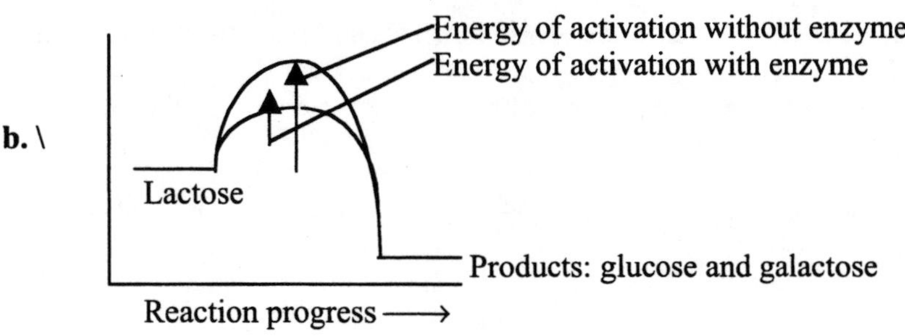

c. By lowering the energy of activation, the enzyme furnishes a lower energy pathway by which the reaction can take place.

21.53 **a.** The disaccharide lactose is a substrate.
b. The *–ase* in lactase indicates that it is an enzyme.
c. The *–ase* in urease indicates that it is an enzyme.
d. Trypsin is an enzyme, which hydrolyzes polypeptides.
e. Pyruvate is a substrate.
f. The *–ase* in transaminase indicates that it is an enzyme.

21.55 **a.** Urea is the substrate of urease.
b. Lactose is the substrate of lactase.
c. Aspartate is the substrate of aspartate transaminase.
d. Tyrosine is the substrate of tyrosine synthetase.

21.57 **a.** The transfer of an acyl group is catalyzed by a transferase.
b. Oxidases are classified as oxidoreductases.
c. A lipase, which splits esters bonds in lipids with water, is a hydrolase.
d. A decarboxylase is classified as a lyase.

21.59 **a.** In this reaction, oxygen is added to an aldehyde. The enzyme that catalyzes this reaction would be an oxidoreductase.
b. In this reaction, a dipeptide is hydrolyzed. The enzyme that catalyzes this reaction would be a hydrolase.
c. In this reaction, water is added to a double bond. The enzyme that catalyzes this reaction would be a lyase.

21.61 Sucrose fits the shape of the active site in sucrase, but lactose does not.

21.63 A heart attack may be the cause. Normally the enzymes LDH and CK are present only in low levels in the blood.

21.65 **a.** An enzyme is saturated if adding more substrate does not increase the rate.
b. An enzyme is unsaturated when increasing the substrate increases the rate.

21.67 In a reversible inhibition, the inhibitor can dissociate from the enzyme, whereas in irreversible inhibition, the inhibitor forms a strong covalent bond with the enzyme and does not dissociate. Irreversible inhibitors act as poisons to enzymes.

21.69 **a.** The oxidation of glycol to an aldehyde and carboxylic acid is catalyzed by an oxidoreductase.
b. At high concentration, ethanol, which acts as a competitive inhibitor of ethylene glycol, would saturate the enzyme to allow ethylene glycol to be removed from the body without producing oxalic acid.

21.71 **a.** Antibiotics such as amoxicillin are irreversible inhibitors.
b. Antibiotics inhibit enzymes needed to form cell walls in bacteria, not humans.

21.73 **a.** When pepsinogen enters the stomach, the low pH cleaves a peptide from its protein chain to form pepsin.
b. An active protease would digest the proteins of the pancreas rather than the proteins in the foods entering the stomach.

21.75 An allosteric enzyme contains sites for regulators that alter the enzyme and speed up or slow down the rate of the catalyzed reaction.

21.77 The end product of the reaction pathway is a negative regulator that binds to the enzyme to decrease or stop the first reaction in the reaction pathway.

21.79 **a.** The Mg^{2+} is a cofactor, which is required by this enzyme.
b. A protein that is catalytically active is a simple enzyme.
c. The folic acid is a coenzyme, which is required by this enzyme.

21.81 **a.** Coenzyme A requires pantothenic acid (B_5).
b. NAD^+ requires niacin (B_3).
c. Biocytin requires biotin.

21.83 A vitamin combines with an enzyme only when the enzyme and coenzyme are needed to catalyze a reaction. When the enzyme is not needed, the vitamin dissociates for use by other enzymes in the cell.

21.85 **a.** A deficiency of niacin can lead to pellagra.
b. A deficiency of vitamin A can lead to night blindness.
c. A deficiency of vitamin D can weaken bone structure.

22

Nucleic Acids

Study Goals

- Draw the structures of the nitrogen bases, sugars, and nucleotides in DNA and RNA.
- Describe the structures of DNA and RNA.
- Explain the process of DNA replication.
- Describe the preparation of recombinant DNA.
- Describe the transcription process during the synthesis of mRNA.
- Use the codons in the genetic code to describe protein synthesis.
- Explain how an alteration in the DNA sequence can lead to mutations in proteins.
- Describe the regulation of protein synthesis in the cells.

Think About It

1. Where is DNA in your cells?

2. How does DNA determine your height, or the color of your hair or eyes?

3. What is the genetic code?

4. How does a mutation occur?

5. What is recombinant DNA?

Key Terms

Match the statements show below with the following key terms:

a. DNA **b.** RNA **c.** double helix **d.** mutation **e.** transcription

1. _____ The formation of mRNA to carry genetic information from DNA to protein synthesis

2. _____ The genetic material containing nucleotides and nitrogenous bases adenine, cytosine, guanine, and thymine

3. _____ The shape of DNA with a sugar-phosphate backbone and base pairs linked in the center

4. _____ A change in the DNA base sequence that may alter the shape and function of a protein

5. _____ A type of nucleic acid with a single strand of nucleotides of adenine, cytosine, guanine, and uracil

Answers **1.** e **2.** a **3.** c **4.** d **5.** b

22.1 Components of Nucleic Acids

- Nucleic acids are composed of four nitrogenous bases, five-carbon sugars, and a phosphate group.
- In DNA, the nitrogen bases are adenine, thymine, guanine, or cytosine. In RNA, uracil replaces thymine.
- In DNA, the sugar is deoxyribose; in RNA the sugar is ribose.

◆ Learning Exercise 22.1A

1. Write the names and abbreviations for the nitrogen bases in each of the following.

DNA ______________________________

RNA ______________________________

2. Write the name of the sugar in each of the following nucleotides.

DNA ______________________________

RNA ______________________________

Answers
1. DNA: adenine (A), thymine (T), guanine (G), cytosine (C)
   RNA: adenine (A), uracil (U), guanine (G), cytosine (C)
2. DNA: deoxyribose RNA: ribose

◆ Learning Exercise 22.1B

Name each of the following and classify it as a purine or a pyrimidine.

a.

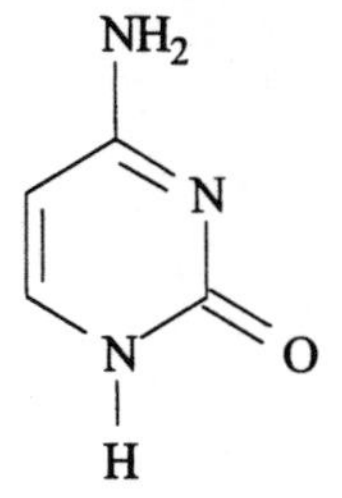

b.

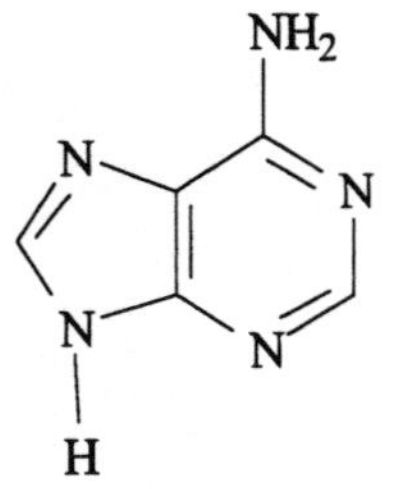

c.

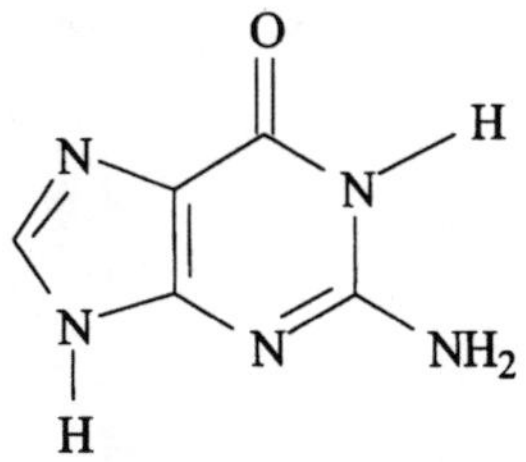

d.

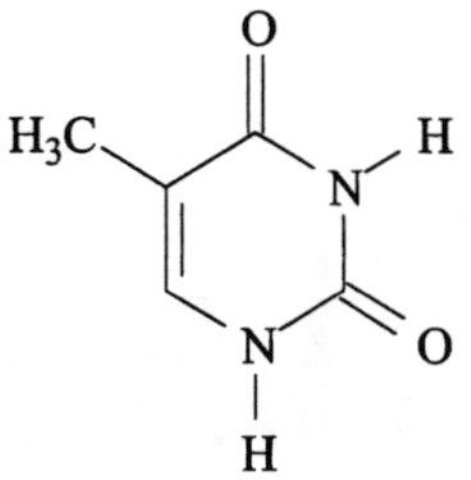

Answers
1. cytosine; pyrimidine
2. adenine; purine
3. guanine; purine
4. thymine, pyrimidine

22.2 Nucleosides and Nucleotides

- A nucleoside is composed of a nitrogen base and a sugar.
- A nucleotide is composed of three parts: a nitrogen base, a sugar, and a phosphate group.
- Deoxyribonucleic acid (DNA) and ribonucleic acid (RNA) are polymers of nucleotides.

◆ Learning Exercise 22.2A

Identify the nucleic acid (DNA or RNA) in which each of the following are found.

1. ____ adenosine-5′-monophosphate

2. _____ dCMP

3. ____ deoxythymidine-5′-monophosphate

4. _____ dGMP

5. ____ guanosine-5′-monophosphate

6. _____ cytidine-5′-monophosphate

7. ____ UMP

8. _____ deoxyadenosine-5′-monophosphate

Answers

1. RNA	**2.** DNA	**3.** DNA	**4.** DNA
5. RNA	**6.** RNA	**7.** RNA	**8.** DNA

◆ Learning Exercise 22.2B

Write the structural formula for deoxyadenosine-5′-monophosphate. Indicate the 5′- and the 3′-carbon atoms on the sugar.

Answer

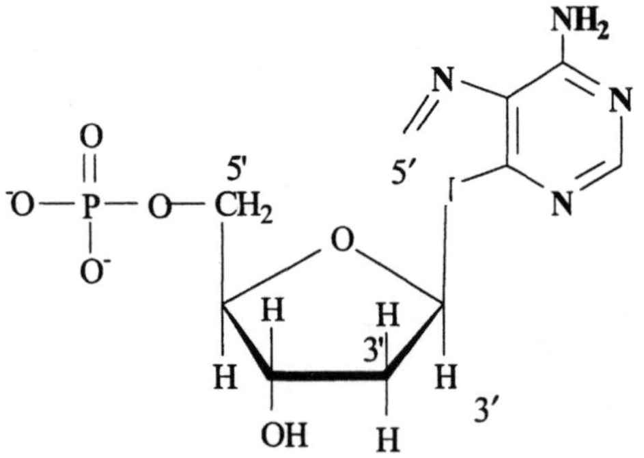

Deoxyadenosine 5'-monophosphate (dAMP)

22.3 Primary Structures of Nucleic Acids

- Nucleic acids are polymers of nucleotides in which the —OH group on the 3′-carbon of a sugar in one nucleotide bonds to the phosphate group attached to the 5′-carbon of a sugar in the adjacent nucleotide.

◆ Learning Exercise 22.3A

In the following dinucleotide, identify each nucleotide, the phosphodiester bond, the 5′-free phosphate group, and the free 3′-hydroxyl group.

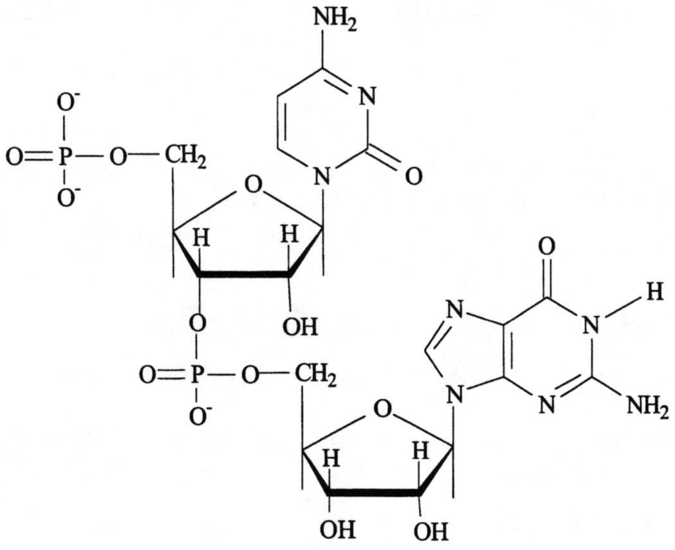

Answer

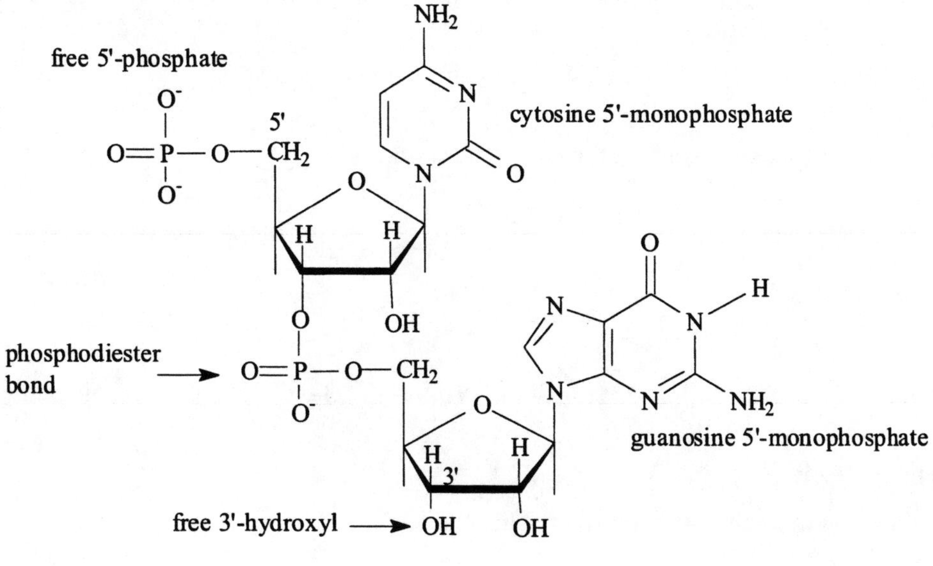

free 5'-phosphate

cytosine 5'-monophosphate

phosphodiester bond

guanosine 5'-monophosphate

free 3'-hydroxyl

◆ Learning Exercise 22.3B

Consider the following sequence of nucleotides in RNA: —A—G—U—C—

1. What are the names of the nucleotides in this sequence?

2. Which nucleotide has the free 5′-phosphate group? ____________________

3. Which nucleotide has the free 3′-hydroxyl group? ____________________

Answer

1. adenosine 5′-monophosphate, guanosine 5′-monophosphate, uridine 5′-monophosphate, cytosine 5′-monophosphate
2. adenosine 5′-monophosphate (AMP) read as 5′—A—G—C—T—3′
3. cytosine 5′-monophosphate (CMP)

22.4 DNA Double Helix: A Secondary Structure

- The two strands in DNA are held together by hydrogen bonds between complementary base pairs, A with T, and G with C.
- One DNA strand runs in the 5′-3′ direction with a free 5′ phosphate, and the other strand runs in the 3′-5′ direction with a free 3′ phosphate.

◆ Learning Exercise 22.4A

Complete the following statements.

1. The structure of the two strands of nucleotides in DNA is called a ____________________.
2. In one strand of DNA, the sugar-phosphate backbone runs in the 5′-3′ direction, whereas the opposite strand goes in the ________________ direction.
3. On the DNA strand that runs in the 5'-3' direction, the free phosphate group is at the ________ end and the free hydroxyl group is at the __________ end.
4. The only combinations of base pairs the connect the two DNA strands are ________________ and ________________.
5. The base pairs along one DNA strand are ____________________________ to the base pairs on the opposite strand.

Answers **1.** double helix **2.** 3′-5′ **3.** 5′-, 3′ **4.** A—T; G—C **5.** complementary

◆ Learning Exercise 22.4B

Complete each DNA section by writing the complementary strand:

1. 5′—ATGCTTGGCTCC—3′
2. 5′—AAATTTCCCGGG—3′
3. 5′—GCGCTCAAATGC—3′

Answers

1. 3′—TACGAACCGAGG—5′
2. 3′—TTTAAAGGGCCC—5′
3. 3′—CGCGAGTTTACG—5′

22.5 DNA Replication

- During DNA replication, DNA polymerase makes new DNA strands along each of the original DNA strands that serve as templates.
- Complementary base pairing ensures the correct pairing of bases to give identical copies of the original DNA.

◆ Learning Exercise 22.5A

How does the replication of DNA produce identical copies of the DNA?

Answer In the replication process, the bases on each strand of the separated parent DNA are paired with their complementary bases. Because each complementary base is specific for a base in DNA, the new DNA strands exactly duplicate the original strands of DNA.

◆ Learning Exercise 22.5B

Match each of the following terms with components or events in DNA replication.

a. replication fork **b.** Okazaki fragment **c.** DNA polymerase
d. helicase **e.** leading strand **f.** lagging strand

1. _____ The enzyme that catalyzes the unwinding of a section of the DNA double helix

2. _____ The points in open sections of DNA where replication begins

3. _____ The enzyme that catalyzes the formation of phosphodiester bonds between nucleotides

4. _____ Short segments produced in the formation of the 3′-5′ DNA daughter strand

5. _____ The new DNA strand that grows in the 5′ to 3′ direction during the formation of daughter DNA

6. _____ The new DNA strand that is synthesized in the 3′ to 5′ direction.

Answers **1.** d **2.** a **3.** c **4.** b **5.** e **6.** f

22.6 Types of RNA

- The three types of RNA differ by function in the cell: ribosomal RNA makes up most of the structure of the ribosomes, messenger RNA carries genetic information from the DNA to the ribosomes, and transfer RNA places the correct amino acids in the protein.

◆ Learning Exercise 22.6

Match each of the following characteristics with a specific type of RNA: mRNA, tRNA, or rRNA.

1. the most abundant type of RNA in a cell _______________

2. the RNA that has the shortest chain of nucleotides _______________

3. the RNA that carries information from DNA to the ribosomes for protein synthesis _______________

4. the RNA that is the major component of ribosomes _______________

5. the RNA that carries specific amino acids to the ribosome for protein synthesis _______________

6. the RNA that consists of a large and a small subunit _______________

Answers
1. rRNA **2.** tRNA **3.** mRNA
4. rRNA **5.** tRNA **6.** rRNA

22.7 Transcription: Synthesis of mRNA

- Transcription is the process by which RNA polymerase produces mRNA from one strand of DNA.
- The bases in the mRNA are complementary to the DNA, except U is paired with A in DNA.
- The polymerase enzyme moves along an unwound section of DNA in a 3' to 5' direction.
- In eukaryotes, initial RNA includes noncoding sections, which are removed before the RNA leaves the nucleus.
- The production of mRNA occurs when certain proteins are needed in the cell.
- In enzyme induction, the appearance of a substrate in a cell removes a repressor, which allows RNA polymerase to produce mRNA at the structural genes.

◆ Learning Exercise 22.7A

Fill in the blanks with a word or phrase that answers each of the following questions:

1. Where in the cell does transcription take place? _______________

2. How many strands of the DNA molecules are involved? _______________

3. Sections in genes that code for proteins _______________

4. Sections in genes that do not code for proteins _______________

5. The abbreviations for the four nucleotides in mRNA are _______________

6. Write the corresponding section of a mRNA produced from each of the following questions.

A. 3′—C—A—T—T—C—G—G—T—A—5′

B. 3′—G—T—A—C—C—T—A—A—C—G—T—C—C—G—5′

Answers
1. nucleus **2.** one **3.** exons **4.** introns **5.** A, U, G, C
6. A. 5′—G—U—A—A—G—C—C—A—U—3′
B. 5′—C—A—U—G—G—A—U—U—G—C—A—G—G—C—3′

◆ Learning Exercise 22.7B

Match the following descriptions of cellular control with the terms.

A. repressor **B.** operon **C.** structural gene
D. enzyme repression **E.** enzyme induction

1. ____ The production of an enzyme caused by the appearance of a substrate
2. ____ A unit formed by a structural gene and an operator gene
3. ____ High levels of an end product stop the production of the enzymes in that pathway
4. ____ A protein that attaches to the operator gene and blocks the synthesis of protein
5. ____ The portion of DNA that produces the mRNA for protein synthesis

Answers **1.** E **2.** B **3.** D **4.** A **5.** C

22.8 The Genetic Code

- The genetic code consists of a sequence of three bases (triplet) that specifies the order for the amino acids in a protein.
- There are 64 codons for the 20 amino acids, which means there are several codons for most amino acids.
- The codon AUG signals the start of transcription and codons UAG, UGA, and UAA signal stop.

◆ Learning Exercise 22.8

Indicate the amino acid coded for by the following mRNA codons.

1. UUU ____________ **2.** GCG ____________
3. AGC ____________ **4.** CCA ____________
5. GGA ____________ **6.** ACA ____________
7. AUG ____________ **8.** CUC ____________
9. CAU ____________ **10.** GUU ____________

Answers **1.** Phe **2.** Ala **3.** Ser **4.** Pro **5.** Gly
6. Thr **7.** Start/Met **8.** Leu **9.** His **10.** Val

22.9 Protein Synthesis: Translation

- Proteins are synthesized at the ribosomes in a translation process that includes three steps: initiation, elongation, and termination.
- During translation, the different tRNA molecules bring the appropriate amino acids to the ribosome where the amino acid is bonded by a peptide bond to the growing peptide chain.
- When the polypeptide is released, it takes on its secondary and tertiary structures to become a functional protein in the cell.

◆ Learning Exercise 22.9A

Match the following components of the translation process with the following.

a. initiation **b.** activation **c.** anticodon **d.** translocation **e.** termination

1. ____ The three bases in each tRNA that complement a codon on the mRNA

2. ____ The combining of an amino acid with a specific tRNA

3. ____ The placement of methionine on the large ribosome

4. ____ The shift of the ribosome from one codon on mRNA to the next

5. ____ The process that occurs when the ribosome reaches a UAA or UGA codon on mRNA

Answers **1.** c **2.** b **3.** a **4.** d **5.** e

◆ Learning Exercise 22.9B

Write the mRNA that would form for the following section of DNA. For each codon in the mRNA, write the amino acid that would be placed in the protein by a tRNA.

1. DNA strand: 3′—CCC—TCA—GGG—CGC—5′

mRNA: ______—______—______—______

amino acid order: __________—________—________—________

2. DNA: 3′—ATA—GCC—TTT—GGC—AAC—5′

mRNA: ______—______—______—______—______

amino acid order: __________—________—________—________—________

Answers

1. mRNA: 5′—GGG—AGU—CCC—GCG—3′
—Gly—Ser—Pro—Ala—

2. mRNA: 5′—UAU—CGG—AAA—CCG—UUG—3′
—Tyr—Arg—Lys—Pro—Leu—

◆ Learning Exercise 22.9C

A segment of DNA that codes for a protein contains 270 nucleic acids. How many amino acids would be present in the protein for this DNA segment?

Answer Assuming that the entire segment codes for a protein, there would be 90 (270 ÷ 3) amino acids in the protein produced.

22.10 Genetic Mutations

- A genetic mutation is a change of one or more bases in the DNA sequence that may alter the structure and ability of the resulting protein to function properly.
- In a substitution, one base is altered, which codes for a different amino acid.
- In a frame shift mutation, the insertion or deletion of one base alters all the codons following the base change, which affects the amino acid sequence that follows the mutation.

◆ Learning Exercise 22.10

Consider the DNA template of 3′—AAT—CCC—GGG—5′.

1. Write the mRNA produced.

   ______—______—______

2. Write the amino acid order for the mRNA codons.

   __________—________—________

3. Suppose a point mutation replaces the thymine in the DNA template with a guanine. Write the mRNA it produces.

   ______—______—______

4. What is the new amino acid order?

   __________—________—________

5. Why is this effect referred to as a point mutation?

6. How is a point mutation different from an insertion or deletion mutation?

7. What are some possible causes of genetic mutations?

Answers

1. 5′—UUA—GGG—CCC—3′ 2. Leu-Gly-Pro
3. 5′—UUC—GGG—CCC— 3′ 4. Phe-Gly-Pro
5. In a point mutation, only one codon is affected, and one amino acid substituted.
6. In an insertion or deletion mutation, the triplet codes that follow the mutation point are shifted by one base, which changes the amino acid order.
7. X-rays, UV light, chemicals called mutagens, and some viruses are possible causes of mutations.

22.11 Recombinant DNA

- Recombinant DNA is DNA that has been synthesized by opening a piece of DNA and inserting a DNA section from another source.
- Much of the work in recombinant DNA is done with the small circular DNA molecules called *plasmids* found in *Escherichia coli* bacteria.
- Recombinant DNA is used to produce large numbers of copies of foreign DNA that is useful in genetic engineering techniques.

◆ Learning Exercise 22.11

Match the statements shown below with the following terms.

A. plasmids **B.** restriction enzymes **C.** polymerase chain reaction
D. recombinant DNA **E.** Human Genome Project

1. _____ A synthetic form of DNA that contains a piece of foreign DNA

2. _____ Research that is mapping the DNA sequences for all the genes in a human cell

3. _____ Small, circular, DNA molecules found in *E. coli* bacteria

4. _____ A process that makes multiple copies of DNA in a short amount of time

5. _____ Enzymes that cut open the DNA strands in the plasmids

Answers **1.** D **2.** E **3.** A **4.** C **5.** B

22.12 Viruses and AIDS

- Viruses are small particles of 3-200 genes that cannot replicate unless they invade a host cell.
- A viral infection involves using the host cell machinery to replicate the viral DNA.
- A retrovirus contains RNA and a reverse transcriptase enzyme that synthesizes a viral DNA in a host cell.

◆ Learning Exercise 22.12

Match the key terms with the statements shown below

A. host cell **B.** retrovirus **C.** vaccine **D.** protease **E.** virus

1. _____ The enzyme inhibited by drugs that prevent the synthesis of viral proteins

2. _____ A small, disease-causing particle that contains either DNA or RNA as its genetic material

3. _____ A type of virus that must use reverse transcriptase to make a viral DNA

4. _____ Required by viruses to replicate

5. _____ Inactive form of viruses that boosts the immune response by causing the body to produce antibodies

Answers **1.** D **2.** E **3.** B **4.** A **5.** C

Check List for Chapter 22

You are ready to take the practice test for Chapter 22. Be sure that you have accomplished the following learning goals for this chapter. If you are not sure, review the section listed at the end of the goal. Then apply your new skills and understanding to the practice test.

After studying Chapter 22, I can successfully:

_____ Identify the components of nucleic acids RNA and DNA (22.1).

_____ Describe the nucleotides contained in DNA and RNA (22.2).

_____ Describe the primary structure of nucleic acids (22.3).

_____ Describe the structures of RNA and DNA; show the relationship between the bases in the double helix (22.4).

_____ Explain the process of DNA replication (22.5)

_____ Describe the structures and characteristics of the three types of RNA (22.6).

_____ Describe the synthesis of mRNA (transcription) (22.7).

_____ the function of the codons in the genetic code (22.8).

_____ Describe the role of translation in protein synthesis (22.9).

_____ Describe some ways in which DNA is altered to cause mutations (22.10).

_____ Describe the process used to prepare recombinant DNA (22.11).

_____ Explain how retroviruses use reverse transcription to synthesize DNA (22.12)

Practice Test for Chapter 22

1. A nucleotide contains

A. a nitrogen base. **B.** a nitrogen base and a sugar.
C. a phosphate and a sugar. **D.** a nitrogen base and a deoxyribose.
E. a nitrogen base, a sugar, and a phosphate.

2. The double helix in DNA is held together by

A. hydrogen bonds. **B.** ester linkages. **C.** peptide bonds.
D. salt bridges. **E.** disulfide bonds.

3. The process of producing DNA in the nucleus is called

A. complementation. **B.** replication. **C.** translation.
D. transcription. **E.** mutation.

4. Which occurs in RNA but **NOT** in DNA?

A. thymine **B.** cytosine **C.** adenine
D. phosphate **E.** uracil

5. Which molecule determines protein structure in protein synthesis?

A. DNA **B.** mRNA **C.** tRNA **D.** rRNA **E.** ribosomes

6. Which type of molecule carries amino acids to the ribosomes?

A. DNA **B.** mRNA **C.** tRNA **D.** rRNA **E.** protein

For questions 7 through 15, select answers from the following nucleic acids.

A. DNA **B.** mRNA **C.** tRNA **D.** rRNA

7. _____ Along with protein, it is a major component of the ribosomes.

8. _____ A double helix consisting of two chains of nucleotides held together by hydrogen bonds between nitrogen bases.

9. _____ A nucleic acid that uses deoxyribose as the sugar.

10. _____ A nucleic acid produced in the nucleus that migrates to the ribosomes to direct the formation of a protein.

11. _____ It can place the proper amino acid into the peptide chain.

12. _____ It has nitrogen bases of adenine, cytosine, guanine, and thymine.

13. _____ It contains the codons for the amino acid order.

14. _____ It contains a triplet called an anticodon loop.

15. _____ This nucleic acid is replicated during cellular division.

For questions 16 through 20, select answers from the following.

A.
```
—A—G—C—C—T—A—
 |  |  |  |  |  |
—T—C—G—G—A—T—
```

B. —A—U—U—G—C—U—C—

C.
```
—A—G—T—U—G—U—
 |  |  |  |  |  |
—T—C—A—A—C—A—
```

D. —G—U—A—

E. —A—T—G—T—A—T—

16. _____ A section of a mRNA

17. _____ An impossible section of DNA

18. _____ A codon

19. _____ A section from a DNA molecul

20. _____ A single strand that would not be possible for mRNA

Read the following statements.

A. tRNA assembles the amino acids at the ribosomes.
B. DNA forms a complementary copy of itself called mRNA.
C. Protein is formed and breaks away.
D. tRNA picks up specific amino acids.
E. mRNA goes to the ribosomes.

Of the statements above, select the order in which they occur during protein synthesis.

21. _____ First step

22. _____ Second step

23. _____ Third step

24. _____ Fourth step

25. _____ Fifth step

For questions 26 through 30, select your answers from the following.

A. mutation **B.** enzyme induction **C.** inducer
D. operon **E.** repressor

26. _____ A unit attaches to the operator gene and blocks the synthesis of a protein.

27. _____ An error in the transmission of the base sequence of DNA

28. _____ A portion of a gene composed of the operating gene and the structural genes

29. _____ A substrate that promotes the synthesis of the enzymes necessary for its metabolism

30. _____ The level of end product regulates the synthesis of the enzymes in that metabolic pathway.

Answers to the Practice Test

1. E	**2.** A	**3.** B	**4.** E	**5.** A
6. C	**7.** D	**8.** A	**9.** A	**10.** B
11. C	**12.** A	**13.** B	**14.** C	**15.** A
16. B, D	**17.** C	**18.** D	**19.** A	**20.** E
21. B	**22.** E	**23.** D	**24.** A	**25.** C
26. E	**27.** A	**28.** D	**29.** B	**30.** C

Answers and Solutions to Selected Text Problems

22.1 DNA contains two purines, adenine (A) and guanine (G) and two pyrimidines, cytosine (C) and thymine (T). RNA contains the same bases, except thymine (T) is replaced by the pyrimidine uracil (U).

a. Pyrimidine **b.** Pyrimidine

22.3 DNA contains two purines, adenine (A) and guanine (G) and two pyrimidines, cytosine (C) and thymine (T). RNA contains the same bases, except thymine (T) is replaced by the pyrimidine uracil (U).

a. DNA **b.** Both DNA and RNA

22.5 Nucleotides contain a base, a sugar, and a phosphate group. The nucleotides found in DNA would all contain the sugar deoxyribose. The four nucleotides are: deoxyadenosine-5′-monophosphate (dAMP), deoxythymidine-5′-monophosphate (dTMP), deoxycytidine-5′-monophosphate (dCMP), and deoxyguanosine-5′-monophosphate (dGMP).

22.7 **a.** Adenosine is a nucleoside found in RNA.
b. Deoxycytidine is a nucleoside found in DNA.
c. Uridine is a nucleoside found in RNA.
d. Cytidine-5′-monophosphate is a nucleotide found in RNA.

22.9

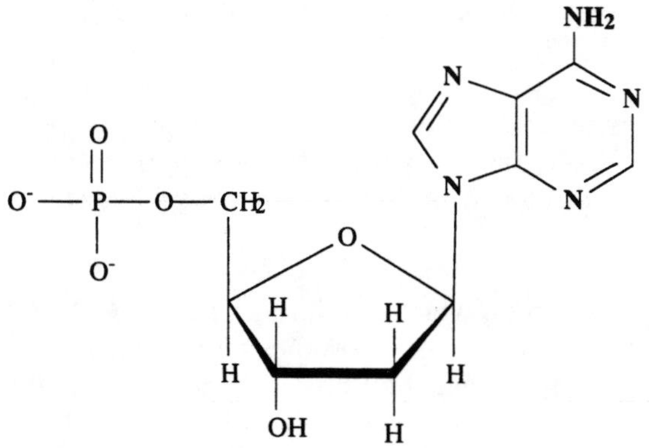

22.11 The nucleotides in nucleic acids are held together by phosphodiester bonds between the 3′ OH of a sugar (ribose or deoxyribose), a phosphate group, and the 5′ OH of another sugar.

22.13

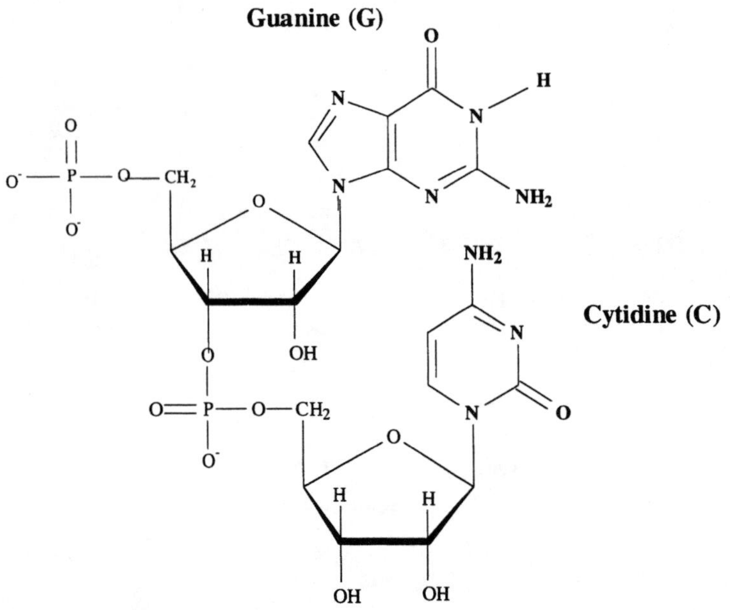

Guanine (G)

Cytidine (C)

22.15 The two DNA strands are held together by hydrogen bonds between the bases in each strand.

22.17 **a.** Since T pairs with A, if one strand of DNA has the sequence 5′—AAAAAA—3′, the second strand would be: 3′—TTTTTT—5′.
b. Since C pairs with G, if one strand of DNA has the sequence 5′—GGGGGG—3′, the second strand would be: 3′—CCCCCC—5′.
c. Since T pairs with A, and C pairs with G, if one strand of DNA has the sequence 5′—AGTCCAGGT—3′, the second strand would be 3′—TCAGGTCCA—5′.
d. Since T pairs with A, and C pairs with G, if one strand of DNA has the sequence 5′—CTGTATACGTTA, the second strand would be: 3′—GACATATGCAAT—5′.

22.19 The enzyme helicase unwinds the DNA helix to prepare the parent DNA strand for the synthesis of daughter DNA strands.

22.21 First, the two DNA strands separate in a way that is similar to the unzipping of a zipper. Then the enzyme DNA polymerase begins to copy each strand by pairing each of the bases in the strands with its complementary base: A pairs with T and C with G. Finally, a phosphodiester bond joins the base to the new, growing strand.

22.23 Ribosomal RNA (rRNA) is found in the ribosomes, which are the sites for protein synthesis. Transfer RNA (tRNA) brings specific amino acids to the ribosomes for protein synthesis. Messenger RNA (mRNA) carries the information needed for protein synthesis from the DNA in the nucleus to the ribosomes.

22.25 A ribosome, which is about 65% rRNA and 35% protein, consists of a small subunit and a large subunit.

22.27 In transcription, the sequence of nucleotides on a DNA template (one strand) is used to produce the base sequences of a messenger RNA. The DNA unwinds and one strand is copied as complementary bases are placed in the mRNA molecule. In RNA, U (uracil) is paired with A in DNA.

22.29 In mRNA, C, G, and A pair with G, C, and T in DNA. However, in mRNA U will pair with A in DNA. The strand of mRNA would have the following sequence:
5′—GGC—UUC—CAA—GUG—3′.

22.31 In eukaryotic cells, genes contain sections called exons that code for protein and sections called introns that do not code for protein.

22.33 An operon is a section of DNA that regulates the synthesis of one or more proteins.

22.35 When the lactose level is low in *E. coli*, a repressor produced by the mRNA from the regulatory gene binds to the operator blocking the synthesis of mRNA from the genes and preventing the synthesis of protein.

22.37 A codon is the three-base sequence (triplet) in mRNA that codes for a specific amino acid in a protein.

22.39 **a.** The codon CUU in mRNA codes for the amino acid leucine.
b. The codon UCA in mRNA codes for the amino acid serine.
c. The codon GGU in mRNA codes for the amino acid glycine.
d. The codon AGG in mRNA codes for the amino acid arginine.

22.41 When AUG is the first codon, it signals the start of protein synthesis and incorporates methionine as the first amino acid in the peptide. Eventually the initial methionine is removed as the protein forms its secondary and tertiary protein structure. In the middle of an mRNA sequence AUG codes for methionine.

22.43 A codon is a base triplet in the mRNA template. An anticodon is the complementary triplet on a tRNA for a specific amino acid.

22.45 The three steps in translation are: initiation, translocation, and termination.

22.47 The mRNA must be divided into triplets and the amino acid coded for by each triplet read from the table.
a. The codon AAA in mRNA codes for lysine: —Lys—Lys—Lys—
b. The codon UUU codes for phenylalanine and CCC for proline: —Phe—Pro—Phe—Pro—
c. —Tyr—Gly—Arg—Cys—

22.49 After a tRNA attaches to the first binding site on the ribosome, its amino acid forms a peptide bond with the amino acid on the tRNA attached to the second binding site. The ribosome moves along the mRNA and a new tRNA with its amino acid occupies the open binding site.

22.51 **a.** By using the pairing: DNA bases C G T A

↓ ↓ ↓ ↓

mRNA bases G C A U

the mRNA sequence can be determined: 5′—CGA—AAA—GUU—UUU—3′.
b. The tRNA triplet anticodons would be as follows: GCU, UUU, CAA, AAA.
c. The mRNA is divided into triplets and the amino acid coded for by each triplet read from the table. Using codons in mRNA: Arg—Lys—Val—Phe.

22.53 In a substitution mutation an incorrect base replaces a base in DNA.

22.55 If the resulting codon still codes for the same amino acid, there is no effect. If the new codon codes for a different amino acid, there is a change in the order of amino acids in the polypeptide.

22.57 The normal triplet TTT in DNA transcribes to AAA in mRNA. AAA codes for lysine. The mutation TTC in DNA transcribes to AAG in mRNA, which also codes for lysine. Thus, there is no effect on protein synthesis.

22.59 **a.** —Thr—Ser—Arg—Val— is the amino acid sequence produced by normal DNA.
b. —Thr—Thr—Arg—Val— is the amino acid sequence produced by a mutation.
c. —Thr—Ser—Gly—Val— is the amino acid sequence produced by a mutation.
d. —Thr—STOP Protein synthesis would terminate early. If this mutation occurs early in the formation of the polypeptide, the resulting protein will probably be nonfunctional.
e. The new protein will contain the sequence —Asp—Ile—Thr—Gly—.
f. The new protein will contain the sequence —His—His—Gly—.

22.61 **a.** Both codons GCC and GCA code for alanine.
b. A vital ionic cross-link in the tertiary structure of hemoglobin cannot be formed when the polar glutamine is replaced by valine, which is nonpolar. The resulting hemoglobin is malformed and less capable of carrying oxygen.

22.63 *E. coli* are used in recombinant DNA work because they are easy, fast, and cheap to grow and they contain plasmids—the portion of the cell into which the foreign DNA can be inserted. Then the plasmids are replicated; as they replicate they make copies of the foreign DNA.

22.65 The cells of *E. coli* are soaked in a detergent solution, which dissolves the plasma membrane and frees the plasmids, which can then be collected.

22.67 A gene for a specific protein is inserted into the plasmid by using a restriction enzyme that cuts the plasmid DNA in specific places. The same enzyme is used to cut a piece from the DNA to be inserted. The cut-out genes and the cut plasmid are mixed, and DNA ligase—an enzyme that catalyzes the joining of DNA—is added. The foreign gene is inserted into the plasmid and E. coli take up the plasmids.

22.69 A DNA fingerprint is the unique set of fragments that are formed from an individual's DNA. Each individual possesses unique DNA and each DNA gives a unique set of fragments and a unique fingerprint.

22.71 A virus contains either DNA or RNA, but not both, inside a protein coating.

22.73 **a.** An RNA-containing virus must make viral DNA from the RNA, a process called reverse transcription.
b. A virus that uses reverse transcription is a retrovirus.

22.75 Nucleoside analogs like AZT or ddI mimic the structures of nucleosides that the HIV virus uses for DNA synthesis. These analogs are incorporated into the new viral DNA chain, but the lack of a hydroxyl group in position 3' of the sugar stops the chain from growing any longer and prevents replication of the virus.

22.77 **a.** pyrimidine **b.** purine **c.** pyrimidine
d. pyrimidine **e.** purine

22.79 **a.** thymine and deoxyribose **b.** adenine and ribose
c. cytosine and ribose **d.** guanine and deoxyribose

22.81 They are both pyrimidines, but thymine has a methyl group.

22.83

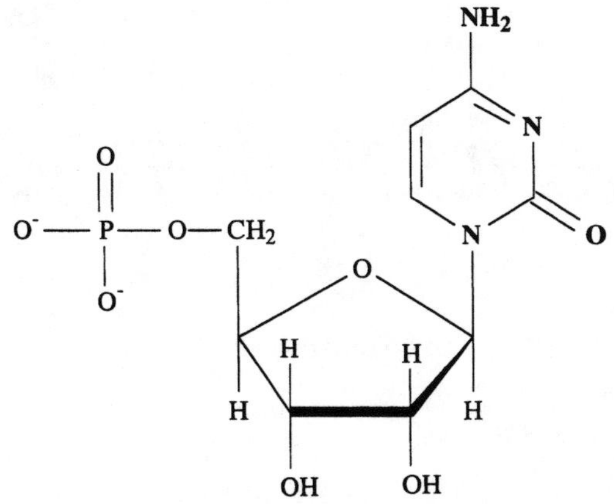

NH_2

N

O

$O^- — P — O — CH_2$

O N O

O^-

H H

H H

OH OH

22.85 They are both polymers of nucleotides connected through phosphodiester bonds between alternating sugar and phosphate groups with bases extending out from each sugar. What is similar about the primary structure of RNA and DNA?

22.87 28% T, 22% G, and 22% C

22.89 **a.** There are two hydrogen bonds between A and T in DNA.
b. There are three hydrogen bonds between G and C in DNA.

22.91 **a.** 3′—CTGAATCCG—5′
b. 5′—ACGTTTGATCGT—3′
c. 3′—TAGCTAGCTAGC—5′

22.93 DNA polymerase synthesizes the leading strand continuously in the 5′ to 3′ direction. The lagging strand is synthesized in small segments called Okazaki fragments because it must grow in the 3′ to 5′ direction.

22.95 One strand of the parent DNA is found in each of the two copies of the daughter DNA molecule.

22.97 **a.** tRNA **b.** rRNA **c.** mRNA

22.99 **a.** ACU, ACC, ACA, and ACG
b. UAU, UCC, UCA, and UCG
c. UGU and UGC

22.101 **a.** AAG codes for lysine **b.** AUU codes for isoleucine **c.** CGG codes for arginine

22.103 Using the genetic code the codons indicate the following:
start(methionine) —Tyr—Gly—Gly—Phe—Leu—stop

22.105 The anticodon consists of the three complementary bases to the codon.
a. UCG **b.** AUA **c.** GGU

22.107 Three nucleotides are needed for each amino acid plus a start and stop triplet, which makes a minimum total of 33 nucleotides.

22.109 A DNA virus attaches to a cell and injects viral DNA that uses the host cell to produce copies of DNA to make viral RNA. A retrovirus injects viral RNA from which complementary DNA is produced by reverse transcription.

23

Metabolic Pathways for Carbohydrates

Study Goals

- Explain the role of ATP in anabolic and catabolic reactions.
- Compare the structures and function of the coenzymes NAD^+, FAD, and coenzyme A.
- Give the sites, enzymes, and products for the digestion of carbohydrates.
- Describe the key reactions in the degradation of glucose in glycolysis.
- Describe the three possible pathways for pyruvate.
- Discuss the impact of ATP levels on glycogen metabolism.
- Describe gluconeogenesis and the Cori cycle.

Think About It

1. Why do you need ATP in your cells?

2. What monosaccharides are produced when carbohydrates undergo digestion?

3. What is meant by *aerobic* and *anaerobic* conditions in the cells?

4. How does glycogen help maintain blood glucose level?

Key Terms

Match the key terms with the correct statement shown below.

a. ATP **b.** glycogen **c.** glycolysis
d. catabolic reaction **e.** mitochondria

1. _____ The storage form of glucose in the muscle and liver

2. _____ A metabolic reaction that produces energy for the cell by degrading large molecules

3. _____ A high-energy compound produced from energy-releasing processes that provides energy for energy-producing reactions

4. _____ The degradation reactions of glucose that yield two pyruvate molecules

5. _____ The organelles in the cells where energy-producing reactions take place

Answers **1.** b **2.** d **3.** a **4.** c **5.** e

23.1 Metabolism and Cell Structure

- Metabolism is all the chemical reactions that provide energy and substances for cell growth.
- Catabolic reactions degrade large molecules to produce energy.
- Anabolic reactions utilize energy in the cells to build large molecules for the cells.
- In cells, different organelles contain the enzymes and coenzymes for the various catabolic and anabolic reactions.

◆ Learning Exercise 23.1A

Match each of the following organelles with their description or cellular function.

a. lysosomes **b.** ribosomes **c.** mitochondria
d. Golgi apparatus **e.** cytoplasm **f.** plasma membrane

1. ____ separates the contents of a cell from the external environment

2. ____ contain enzymes that catalyze energy-producing reactions

3. ____ the cellular material between the plasma membrane and the nucleus

4. ____ modifies proteins from the endoplasmic reticulum for cell membranes

5. ____ protein synthesis

6. ____ hydrolytic enzymes digest old cell structures

Answers **1.** f **2.** c **3.** e **4.** d **5.** b **6.** a

◆ Learning Exercise 23.1B

Identify the stages of metabolism for each of the following processes.

a. Stage 1 **b.** Stage 2 **c.** Stage 3

1. ____ Oxidation of two-carbon acetyl CoA enters a series of reactions that provide most of the energy for ATP synthesis.

2. ____ Polysaccharides undergo digestion to monosaccharides such as glucose.

3. ____ Digestion products such as glucose are degraded to two- or three-carbon compounds.

Answers **1.** c **2.** a **3.** b

23.2 ATP and Energy

- Energy is stored in ATP, a high-energy compound that is hydrolyzed when energy is required for the anabolic reactions that do work in the cells.
- The hydrolysis of ATP, which releases energy, is linked with many anabolic reactions in the cell.

◆ **Learning Exercise 23.2**

Complete the following statements for ATP.

The ATP molecule is composed of a nitrogen-base (1)________________, a (2) ____________ sugar, and three (3) __________________________. ATP undergoes (4)______________, which cleaves a (5) ________________ and releases (6) ________________. For this reason, ATP is called a (7)____________________ compound. The resulting phosphate group, called inorganic phosphate, is abbreviated as (8) ____________. This equation can be written as (9)____. The energy from ATP is linked to cellular reactions that are (10)____________________________.

Answers

1. adenine **2.** ribose **3.** phosphate groups **4.** hydrolysis
5. phosphate **6.** energy **7.** high-energy **8.** P_i
9. $ATP + H_2O \longrightarrow ADP + P_i + \text{Energy (7.3 kcal/mole)}$ **10.** energy requiring

23.3 Important Coenzymes in Metabolic Pathways

- Coenzymes such as FAD and NAD^+ pick up hydrogen and electrons during oxidative processes.
- Coenzyme A is a coenzyme that carries acetyl (two-carbon) groups produced when glucose, fatty acids, and amino acids are degraded.

Select the coenzyme that matches each of the following descriptions of coenzymes.

a. NAD^+ **b.** NADH **c.** FAD **d.** $FADH_2$ **e.** coenzyme A

1. ____ participates in reactions that convert a hydroxyl group to a C=O group

2. ____ contains riboflavin (vitamin B_2)

3. ____ reduced form of nicotinamide adenine dinucleotide

4. ____ contains the vitamin niacin

5. ____ oxidized form of flavin adenine dinucleotide

6. ____ contains the vitamin pantothenic acid ADP, and an aminoethanethiol

7. ____ participates in oxidation reactions that produce a carbon–carbon double bond (C=C)

8. ____ transfers acyl groups such as the two-carbon acetyl group

9. ____ reduced form of flavin adenine dinucleotide

Answers

1. a **2.** c, d **3.** b **4.** a, b **5.** c
6. e **7.** c **8.** e **9.** d

23.4 Digestion of Carbohydrates

- Digestion is a series of reactions that break down large food molecules of carbohydrates, lipids, and proteins into smaller molecules that can be absorbed and used by the cells.
- The end products of digestion of polysaccharides are the monosaccharides glucose, fructose, and galactose.

◆ Learning Exercise 23.4

Complete the table to describe sites, enzymes, and products for the digestion of carbohydrates.

Food	Digestion site(s)	Enzyme	Products
1. amylose			
2. amylopectin			
3. maltose			
4. lactose			
5. sucrose			

Answers

Food	Digestion site(s)	Enzyme	Products
1. amylose	**a.** mouth **b.** small intestine (mucosa)	**a.** salivary amylase **b.** pancreatic amylase	**a.** smaller polysaccharides (dextrins), some maltose and glucose **b.** maltose, glucose
2. amylopectin	**a.** mouth **b.** small intestine (mucosa)	**a.** salivary amylase **b.** pancreatic amylase, branching enzyme	**a.** smaller polysaccharides (dextrins), some maltose and glucose **b.** maltose, glucose
3. maltose	small intestine	maltase	glucose and glucose
4. lactose	small intestine	lactase	glucose and galactose
5. sucrose	small intestine	sucrase	glucose and fructose

23.5 Glycolysis: Oxidation of Glucose

- Glycolysis is the primary anaerobic pathway for the degradation of glucose to yield pyruvic acid.
- Glucose is converted to fructose-1,6-diphosphate that is split into two triose phosphate molecules.
- The oxidation of the three-carbon sugars yields the reduced coenzyme 2 NADH and 2 ATP.

◆ Learning Exercise 23.5

Match each of the following terms of glycolysis with the best description.

a. 2 NADH **b.** anaerobic **c.** glucose **d.** two pyruvate
e. energy invested **f.** energy generated **g.** 2 ATP **h.** 4 ATP

1. ____ the starting material for glycolysis **2.** ____ Steps 1–5 of glycolysis

3. ____ operates without oxygen **4.** ____ net ATP energy produced

5. ____ number of reduced coenzymes produced **6.** ____ Steps 6–10 of glycolysis

7. ____ end product of glycolysis **8.** ____ number of ATP required

Answers **1.** c **2.** e **3.** b **4.** g **5.** a
6. f **7.** d, a **8.** g

23.6 Pathways for Pyruvate

- In the absence of oxygen, pyruvate is reduced to lactate and NAD^+ is regenerated for the continuation of glycolysis.
- Under aerobic conditions, pyruvate is oxidized in the mitochondria to acetyl CoA, which enters the citric acid cycle.

◆ Learning Exercise 23.6A

Fill in the blanks with the following terms.

lactate NAD^+ fermentation
NADH aerobic anaerobic acetyl CoA

When oxygen is available during glycolysis, the three-carbon pyruvate may be oxidized to form (1)____________________ + CO_2. The coenzyme (2) ______ is reduced to (3)__________. Under (4)_______________ conditions, pyruvate is reduced to (5) _______________. In yeast, pyruvate forms ethanol in a process known as (6)____________________.

Answers **1.** acetyl CoA **2.** NAD^+ **3.** NADH
4. anaerobic **5.** lactate **6.** fermentation

◆ Learning Exercise 23.6B

Essay: Explain how the formation of lactate from pyruvate during anaerobic conditions allows glycolysis to continue.

Answer Under anaerobic conditions, the oxidation of pyruvate to acetyl CoA to regenerate NAD^+ cannot take place. Then, pyruvate is reduced to lactate using NADH in the cytoplasm and regenerating NAD^+.

23.7 Glycogen Metabolism

◆ **Learning Exercise 23.7**

Associate each of the following descriptions with pathways in glycogen metabolism.

a. glycogenesis **b.** glycogenolysis

1. ____ break down of glycogen to glucose **2.** ____ activated by glucagon

3. ____ starting material is glucose-6-phosphate **4.** ____ synthesis of glycogen from glucose

5. ____ activated by insulin **6.** ____ UDP activates glucose

Answers **1.** b **2.** b **3.** a **4.** a **5.** a **6.** a

23.8 Gluconeogenesis: Glucose Synthesis

◆ **Learning Exercise 23.7**

Associate each of the following descriptions.

a. gluconeogenesis **b.** pyruvate **c.** pyruvate kinase
d. pyruvate carboxylase **e.** Cori cycle

1. ____ an enzyme in glycolysis that cannot be utilized in gluconeogenesis

2. ____ a typical noncarbohydrate source of carbon atoms for glucose synthesis

3. ____ a process whereby lactate produced in muscle is used for glucose synthesis in the liver and used again by the muscle

4. ____ the metabolic pathway that converts noncarbohydrate sources to glucose

5. ____ an enzyme used in gluconeogenesis that is not used in glycolysis

6. ____ a metabolic pathway that is activated when glycogen reserves are depleted

Answers **1.** c **2.** b **3.** e **4.** a **5.** d **6.** a

Check List for Chapter 23

You are ready to take the practice test for Chapter 23. Be sure that you have accomplished the following learning goals for this chapter. If you are not sure, review the section listed at the end of the goal. Then apply your new skills and understanding to the practice test.

After studying Chapter 23 I can successfully:

____ Associate catabolic and anabolic reactions with organelles in the cell (23.1).

____ Describe the role of ATP in catabolic and anabolic reactions (23.2).

____ Describe the coenzymes NAD^+, FAD, and Coenzyme A (23.3).

____ Describe the sites, enzymes, and products of digestion for carbohydrates (23.4).

____ Describe the conversion of glucose to pyruvate in glycolysis (23.5).

____ Give the conditions for the conversion of pyruvate to lactate, ethanol, and acetyl coenzyme A (23.6).

____ Describe the formation and breakdown of glycogen (23.7).

____ Describe the reactions in which noncarbohydrate sources are used to synthesize glucose (23.8).

Practice Test for Chapter 23

1. The main function of the mitochondria is

A. energy production. **B.** protein synthesis. **C.** glycolysis.
D. genetic instructions. **E.** waste disposal.

2. ATP is a(n)

A. nucleotide unit in RNA and DNA
B. end product of glycogenolysis
C. end product of transamination
D. enzyme
E. energy storage molecule

Use one or more of the following enzymes and end products for the digestion of each of items 3 through 6.

A. maltase **B.** glucose **C.** fructose **D.** sucrase

E. galactose **F.** lactase **G.** pancreatic amylase

3. sucrose ____________ **4.** lactose ____________

5. small polysaccharides ____________ **6.** maltose ____________

Match the parts of the cell with each of the following descriptions.

A. mitochondria **B.** lyosomes **C.** cytosol
D. ribosomes **E.** plasma membrane

7. ____ protein synthesis **8.** ____ fluid part of the cytoplasm

9. ____ separates cell contents from external fluids **10.** ____ energy-producing reactions

11. ____ hydrolytic enzymes degrade old cell structures

12. Glycolysis

A. requires oxygen for the catabolism of glucose.
B. represents the aerobic sequence for glucose anabolism and ATP production.
C. represents the splitting off of glucose residues from glycogen.
D. represents the anaerobic catabolism of glucose to pyruvate.
E. produces acetyl units and ATP as end products.

13. Which does **NOT** appear in the glycolysis pathway?

A. dihydroxyacetone phosphate **B.** pyruvate **C.** NAD^+
D. acetyl CoA **E.** lactate

Match each of the following with the correct metabolic pathway

A. glycolysis **B.** glycogenolysis **C.** gluconeogenesis
D. glycogenesis **E.** fermentation

14. ____ conversion of pyruvate to alcohol **15.** ____ breakdown of glucose to pyruvate

16. ____ formation of glycogen **17.** ____ synthesis of glucose

18. ____ break down of glycogen to glucose

Associate each of the following coenzymes with the correct description.

A. NAD^+ **B.** NADH **C.** FAD **D.** $FADH_2$ **E.** coenzyme A

19. ____ converts a hydroxyl group to a C=O group

20. ____ reduced form of nicotinamide adenine dinucleotide

21. ____ oxidized form of flavin adenine dinucleotide

22. ____ contains the vitamin pantothenic acid ADP, and an aminoethanethiol

23. ____ participates in oxidation reactions that produce a carbon–carbon double bond (C=C)

24. ____ transfers acyl groups such as the two-carbon acetyl group

25. ____ reduced form of flavin adenine dinucleotide

Answer the following questions for glycolysis.

26. ____ Number of ATP invested to oxidation one glucose molecule

27. ____ Number of ATP (net) produced from one glucose molecule

28. ____ Number of NADH produced from the degradation of one glucose molecule

Answers to the Practice Test

1. A	**2.** E	**3.** B, C, D	**4.** B, E, F	**5.** G, B
6. A, B	**7.** D	**8.** C	**9.** E	**10.** A
11. B	**12.** D	**13.** D	**14.** E	**15.** A
16. D	**17.** C	**18.** B	**19.** A	**20.** B
21. C	**22.** E	**23.** C	**24.** E	**25.** D
26. 2	**27.** 2	**28.** 2		

Answers and Solutions to Selected Text Problems

23.1 The digestion of polysaccharides takes place in stage 1.

23.3 A catabolic reaction breaks down larger molecules to smaller molecules accompanied by the release of energy.

23.5
a. (3) Smooth endoplasmic reticulum is the site for the synthesis of fats and steroids.
b. (1) Lysosomes contain hydrolytic enzymes.
c. (2) The Golgi complex modifies products from the rough endoplasmic reticulum.

23.7 The phosphoric anhydride bonds (P—O—P) in ATP release energy that is sufficient for energy-requiring processes in the cell.

23.9
a. $\text{PEP} + H_2O \longrightarrow \text{pyruvate} + P_i + 14.8\ \text{kcal/mole}$
b. $\text{ADP} + P_i + 7.3\ \text{kcal/mole} \longrightarrow \text{ATP} + H_2O$
c. Coupled: $\text{PEP} + \text{ADP} \longrightarrow \text{ATP} + \text{pyruvate} + 7.5\ \text{kcal/mole}$

23.11 **a.** Pantothenic acid is a component in coenzyme A.
b. Niacin is the vitamin component of NAD^+.
c. Ribitol is the alcohol sugar that makes up riboflavin in FAD.

23.13 In biochemical systems, oxidation is usually accompanied by gain of oxygen or loss of hydrogen. Loss of oxygen or gain of hydrogen usually accompanies reduction.

a. The reduced form of NAD^+ is abbreviated NADH.
b. The oxidized form of $FADH_2$ is abbreviated FAD.

23.15 The coenzyme FAD accepts hydrogen when a dehydrogenase forms a carbon–carbon double bond.

23.17 Digestion breaks down the large molecules in food into smaller compounds that can be absorbed by the body. Hydrolysis is the main reaction involved in the digestion of carbohydrates.

23.19 **a.** The disaccharide lactose is digested in the small intestine to yield galactose and glucose.
b. The disaccharide sucrose is digested in the small intestine to yield glucose and fructose.
c. The disaccharide maltose is digested in the small intestine to yield two glucose.

23.21 Glucose is the starting reactant for glycolysis.

23.23 In the initial reactions of glycolysis, two ATP molecules are required to add phosphate groups to the glucose.

23.25 When fructose-1,6-bisphosphate splits, glyceraldehyde-3-phosphate and dihydroxyacetone phosphate are formed. The dihydroxyacetone phosphate is converted to glyceraldehyde-3-phosphate for subsequent reactions.

23.27 ATP is produced directly in glycolysis in two places. In reaction 7, phosphate from 1,3-bisphosphoglycerate is transferred to ADP and to yield ATP. In reaction 10, phosphate from phosphoenolpyruvate is transferred directly to ADP.

23.29 **a.** In glycolysis, phosphorylation is catalyzed by the enzyme hexokinase.
b. In glycolysis, direct transfer of a phosphate group is catalyzed by the enzyme phosphokinase.

23.31 **a.** In the phosphorylation of glucose to glucose-6-phosphate 1 ATP is required.
b. One ATP is required for the conversion of glyceraldehyde-3-phosphate to 1,3-bisphosphoglycerate.
c. When glucose is converted to pyruvate, two ATP and two NADH are produced.

23.33 **a.** The first ATP is hydrolyzed in the first reaction in glycolysis; the change of glucose to glucose-6-phosphate.
b. Direct substrate phosphorylation occurs in reaction 7 of glycolysis when the transfer of phosphate from 1,3-bisphosphoglycerate to ADP generates ATP. In reaction 10 of glycolysis, phosphate is transferred from phosphoenolpyruvate directly to ADP.
c. In reaction 4 of glycolysis, the six-carbon fructose-1,6-bisphosphate is converted to two three-carbon molecules.

23.35 Galactose reacts with ATP to yield galactose-1-phosphate, which is converted to glucose phosphate, an intermediate in glycolysis. Fructose reacts with ATP to yield fructose-1-phosphate, which is cleaved to give dihydroxyacetone phosphate and glyceraldehyde. Dihydroxyacetone phosphate isomerizes to glyceraldehyde-3-phosphate, and glyceraldehyde is phosphorylated to glyceraldehyde-3-phosphate, which is an intermediate in glycolysis.

23.37 **a.** Low levels of ATP activate phosphofructokinase to increase the rate of glycolysis.
b. When ATP levels are high, ATP inhibits phosphofructokinase and slows or prevents glycolysis.

23.39 A cell converts pyruvate to acetyl CoA only under aerobic conditions; there must be sufficient oxygen available.

23.41 The overall reaction for the conversion of pyruvate to acetyl CoA is:

$$\underset{\textit{pyruvate}}{CH_3{-}\overset{\overset{\displaystyle O}{\|}}{C}{-}COO^-} + NAD^+ + HS{-}CoA \longrightarrow \underset{\textit{acetyl CoA}}{CH_3{-}\overset{\overset{\displaystyle O}{\|}}{C}{-}S{-}CoA} + CO_2 + NADH + H^+$$

23.43 The reduction of pyruvate to lactate regenerates NAD^+, which allows glycolysis to proceed and produces a small amount of ATP.

23.45 During fermentation the three-carbon compound pyruvate is reduced to ethanol while decarboxylation removes one carbon as CO_2.

23.47 In glycogenesis, glycogen is synthesized from glucose molecules.

23.49 Muscle cells break down glycogen to glucose 6-phosphate, which enters glycolysis.

23.51 Glycogen phosphorylase cleaves the glycosidic bonds at the ends of glycogen chains to remove glucose monomers as glucose 1-phosphate.

23.53 When there are no glycogen stores remaining in the liver, gluconeogenesis synthesizes glucose from noncarbohydrate compounds such as pyruvate and lactate.

23.55 The enzymes in glycolysis that are also used in their reverse directions for gluconeogenesis are phosphoglucoisomerase, aldolase, triosephosphate isomerase, glyceraldehyde 3-phoshpate dehydrogenase, phosphoglycerokinase, phosphoglyceromutase, and enolase.

23.57 **a.** Low glucose levels activate glucose synthesis.
b. Glucagon produced when glucose levels are low activates gluconeogenesis.
c. Insulin produced when glucose levels are high inhibits gluconeogenesis.

23.59 Metabolism includes all the reactions in cells that provide energy and material for cell growth.

23.61 Stage 1 involves the degradation of large molecules such as polysaccharides.

23.63 A eukaryotic cell has a nucleus, whereas a prokaryotic cell does not.

23.65 ATP is the abbreviation for adenosine triphosphate.

23.67 $ATP + H_2O \longrightarrow ADP + P_i + 7.3$ kcal (31kJ)/mole

23.69 FAD is the abbreviation for flavin adenine dinucleotide.

23.71 NAD^+ is the abbreviation for nicotinamide adenine dinucleotide.

23.73 The reduced forms of these coenzymes include hydrogen obtained from an oxidation reaction.
a. $FADH_2$ **b.** $NADH + H^+$

23.75 Lactose undergoes digestion in the mucosal cells of the small intestine to yield galactose and glucose.

23.77 Galactose and fructose are converted in the liver to glucose phosphate compounds that can enter the glycolysis pathway.

23.79 Glucose is the reactant and pyruvate is the product of glycolysis.

23.81 Reactions 1 and 3 involve phosphorylation of hexoses with ATP, and reactions 7 and 10 involve direct substrate phosphorylation that generates ATP.

23.83 Reaction 4 catalyzed by aldolase converts fructose 1,6-bisphosphate into two triose phosphates.

23.85 Phosphoglucoisomerase converts glucose-6-phosphate to the isomer fructose-6-phosphate.

23.87 Pyruvate is converted to lactate when oxygen is not present in the cell (anaerobic) to regenerate NAD^+ for glycolysis.

23.89 Phosphofructokinase is an allosteric enzyme that is activated by high levels of AMP and ADP because the cell needs to produce more ATP. When ATP levels are high, ATP inhibits phosphofructokinase, which reduces its catalysis of fructose 6-phosphate.

23.91 The rate of glycogenolysis increases when blood glucose levels are low and glucagon has been secreted, which accelerate the breakdown of glycogen.

23.93 The breakdown of glycogen in the liver produces glucose.

23.95 The cells in the liver, but not skeletal muscle, contain a phosphatase enzyme needed to convert glucose 6-phosphate to free glucose that can diffuse through cell membranes into the blood stream. Glucose 6-phosphate, which is the end product of glycogenolysis in muscle cells, cannot diffuse easily across cell membranes.

23.97 Insulin increases the rate of glycogenolysis and glycolysis, and decreases the rate of glycogenesis. Glucagon decreases the rate of glycogenolysis and glycolysis, and increases the rate of glycogenesis.

23.99 The Cori cycle is a cyclic process that involves the transfer of lactate from muscle to the liver where glucose is synthesized, which can be used again by the muscle.

23.101 **a.** Low glucose increases the breakdown of glycogen.
b. Insulin produced when glucose levels are high increases the rate of glycogenesis in the liver.
c. Glucagon secreted when glucose levels are low increases the breakdown of glycogen.
d. High levels of ATP decrease the breakdown of glycogen.

23.103 **a.** High glucose levels decrease the synthesis of glucose.
b. Insulin produced when glucose levels are high decreases glucose synthesis.
c. Glucagon secreted when glucose levels are low increases glucose synthesis.
d. High levels of ATP decrease glucose synthesis.

24

Metabolism and Energy Production

Study Goals

- Describe the reactions in the citric acid cycle that oxidize acetyl CoA.
- Explain how electrons from NADH and H^+ and FAD move along the electron transport chain to form H_2O.
- Describe the role of oxidative phosphorylation in ATP synthesis.
- Calculate the ATP produced by the complete combustion of glucose.

Think About It

1. Why is the citric acid cycle considered a central pathway in metabolism?

2. How is the citric acid cycle connected to electron transport?

3. Which stage of metabolism produces most of the ATP for the cells?

Key Terms

Match the key terms with the correct statement shown below.

a. citric acid cycle	**b.** oxidative phosphorylation	**c.** coenyzme Q
d. cytochromes	**e.** proton pump	**f.** tricarboxylic acid cycle

1. _____ A mobile carrier that passes electrons from NADH and $FADH_2$ to cytochrome b in complex III

2. _____ Proteins containing iron as Fe^{3+} or Fe^{2+} that transfer electrons from QH_2 to oxygen

3. _____ A function of complexes I, II, and III whereby protons move from the matrix into the intermembrane space to create a proton gradient

4. _____ The synthesis of ATP from ADP and P_i using energy generated from electron transport

5. _____ Oxidation reactions that convert acetyl CoA to CO_2 producing reduced coenzymes for energy production via the electron chain transport system.

6. _____ Another name for the citric acid cycle

Answers **1.** c **2.** d **3.** e **4.** b **5.** a **6.** f

24.1 The Citric Acid Cycle

- Under aerobic conditions, pyruvate is oxidized in the mitochondria to acetyl CoA, which enters the citric acid cycle.
- In a sequence of reactions called the citric acid cycle, acetyl CoA combines with oxaloacetate to yield citrate.
- In one turn of the citric acid cycle, the oxidation of acetyl CoA yields two CO_2, GTP, three NADH, and $FADH_2$. The phosphorylation of ADP by GTP yields ATP.

◆ Learning Exercise 24.1A

Match the name of the enzymes with the following steps in the citric acid cycle.

a. isocitrate dehydrogenase **b.** α-ketoglutarate dehydrogenase **c.** fumarase
d. succinate dehydrogenase **e.** malate dehydrogenase **f.** aconitase
g. succinyl CoA synthetase **h.** citrate synthase

1. ____ acetyl CoA + oxaloacetate ⟶ citrate
2. ____ citrate ⟶ isocitrate
3. ____ isocitrate ⟶ α-ketoglutarate acid
4. ____ α-ketoglutarate ⟶ succinyl CoA
5. ____ succinyl CoA ⟶ succinate
6. ____ succinate ⟶ fumarate
7. ____ fumarate ⟶ malate
8. ____ malate ⟶ oxaloacetate
9. ____ allosteric enzymes that regulate the citric acid cycle

Answers **1.** h **2.** f **3.** a **4.** b **5.** g
6. d **7.** c **8.** e **9.** a, b

◆ Learning Exercise 24.1B

In each of the following steps of the citric acid cycle, indicate if oxidation occurs (yes/no) and any coenzyme or direct phosphorylation product produced (NADH + H^+, $FADH_2$, GTP).

Step in citric acid cycle	Oxidation	Coenzyme
1. acetyl CoA + oxaloacetate ⟶ citrate	________	________
2. citrate ⟶ isocitrate	________	________
3. isocitrate ⟶ α-ketoglutarate acid	________	________
4. α-ketoglutarate ⟶ succinyl CoA	________	________
5. succinyl CoA ⟶ succinate	________	________
6. succinate ⟶ fumarate	________	________
7. fumarate ⟶ malate	________	________
8. malate ⟶ oxaloacetate	________	________

Answers **1.** no **2.** no **3.** yes, NADH + H^+ **4.** yes, NADH + H^+
5. no, GTP **6.** yes, $FADH_2$ **7.** no **8.** yes, NADH + H^+

24.2 Electron Carriers

- The reduced coenzymes from glycolysis and the citric acid cycle are oxidized to NAD^+ and FAD by transferring protons and electrons to the electron transport chain.

◆ Learning Exercise 24.2

Write the oxidized and reduced forms of each of the following electron carriers.

1. flavin mononucleotide oxidized ____ reduced ____
2. coenzyme Q oxidized ____ reduced ____
3. iron-protein clusters oxidized ____ reduced ____
4. cytochrome b oxidized ____ reduced ____

Answers

1. oxidized: FMN — reduced: $FMNH_2$
2. oxidized: Q — reduced: QH_2
3. oxidized: Fe^{3+} S cluster — reduced Fe^{2+} S cluster
4. oxidized: Cyt b (Fe^{3+}) — reduced Cyt b (Fe^{2+})

24.3 Electron Transport

- In the electron transport system or respiratory chain, electrons are transferred to electron carriers including flavins, coenzyme Q, iron-sulfur proteins, and cytochromes with Fe^{3+}/Fe^{2+}.
- The final acceptor, O_2, combines with protons and electrons to yield H_2O.

◆ Learning Exercise 24.3A

Identify the protein complexes and mobile carriers in the electron transport chain.

a. cytochrome *c* oxidase (IV) **b.** NADH dehydrogenase (I) **c.** cytochrome *c*
d. coenzyme Q-cytochrome *c* reductase (III) **e.** succinate dehydrogenase (II) **f.** Q

1. ____ FMN and Fe-S clusters
2. ____ $FADH_2 + Q \longrightarrow FAD + QH_2$
3. ____ mobile carrier from complex I or complex II to complex III
4. ____ electrons are passed from cyt *a* and a_3 to O_2 and $4H^+$ to yield $2H_2O$
5. ____ mobile carrier between complex III and IV
6. ____ cyt b and Fe-S clusters

Answers 1. b 2. e 3. f 4. a 5. c 6. d

◆ Learning Exercise 24.3B

1. Write an equation for the transfer of hydrogen from $FMNH_2$ to Q.

2. What is the function of coenzyme Q in the electron transport chain?

3. What are the end products of the electron transport chain?

Answers

1. $FMNH_2 + Q \longrightarrow FMN + QH_2$
2. Q accepts hydrogen atoms from $FMNH_2$ or $FADH_2$. From QH_2, the hydrogen atoms are separated into protons and electrons with the electrons being passed on to the cytochromes, the electron acceptors in the chain.
3. $CO_2 + H_2O$

24.4 Oxidative Phosphorylation and ATP

- The flow of electrons along the electron chain pumps protons across the inner membrane, which produces a high-energy proton gradient that provides energy for the synthesis of ATP.
- The process of using the energy of the electron transport chain to synthesize ATP is called oxidative phosphorylation.

◆ Learning Exercise 24.4

Match the following terms with the correct description below:

a. oxidative phosphorylation **b.** tight (T) site **c.** ATP synthase
d. proton pumps **e.** loose (L) site **f.** open (O) site
g. proton gradient

1. _____ the complexes I, III, and IV through which H^+ ions move out of the matrix into the inner membrane space

2. _____ the protein tunnel where proton flow from the intermembrane space back to the matrix generates energy for ATP synthesis

3. _____ energy from electron transport is used to form a proton gradient that drives ATP synthesis

4. _____ the conformation on the F_1 section of ATP synthase that binds ADP and P_i

5. _____ the accumulation of protons in the intermembrane space that lowers pH

6. _____ the conformation on the F_1 section of ATP synthase that releases ATP

7. _____ the conformation on the F_1 section of ATP synthase where ATP forms

Answers **1.** d **2.** c **3.** a **4.** e **5.** g **6.** f **7.** b

24.5 ATP Energy from Glucose

- The oxidation of NADH yields three ATP molecules, and $FADH_2$ yields two ATP.
- The complete oxidation of glucose yields a total of 36 ATP from direct phosphorylation and the oxidation of the reduced coenzymes NADH and $FADH_2$ by the electron transport chain and oxidative phosphorylation.

◆ Learning Exercise 24.5

Complete the following:

Substrate	Reaction	Products	Amount of ATP
1. glucose	glycolysis (aerobic)		
2. pyruvate	oxidation		
3. acetyl CoA	citric acid cycle		
4. glucose	glycolysis (anaerobic)		
5. glucose	complete oxidation		

Answers

Substrate	Reaction	Products	Amount of ATP
1. glucose	glycolysis (aerobic)	2 pyruvate acid	6 ATP
2. pyruvate acid	oxidation	acetyl CoA + CoA	3 ATP
3. acetyl CoA	citric acid cycle	$2CO_2$	12 ATP
4. glucose	glycolysis (anaerobic)	2 lactate	2 ATP
5. glucose	complete oxidation	$6\ CO_2 + 6\ H_2O$	36 ATP

Check List for Chapter 24

You are ready to take the practice test for Chapter 24. Be sure that you have accomplished the following learning goals for this chapter. If you are not sure, review the section listed at the end of the goal. Then apply your new skills and understanding to the practice test.

After studying Chapter 24 I can successfully:

_____ Describe the oxidation of acetyl CoA in the citric acid cycle (24.1).

_____ Identify the electron carriers in the electron transport system (24.2).

_____ Describe the process of electron transport (24.3).

_____ Explain the chemiosmotic theory whereby ATP synthesis is linked to the energy of the electron transport chain and a proton gradient (24.4).

_____ Account for the ATP produced by the complete oxidation of glucose (24.5).

Practice Test for Chapter 24

1. Which is true of the citric acid cycle?

A. Acetyl CoA is converted to CO_2 and H_2O.
B. Oxaloacetate combines with acetyl units to form citric acid.
C. The coenzymes are NAD^+ and FAD.
D. ATP is produced by direct phosphorylation.
E. all of the above

Match the types of reactions with each of the following.

A. malate **B.** fumarate **C.** succinate
D. citrate **E.** oxaloacetate

2. _____ formed when oxaloacetate combines with acetyl CoA

3. _____ H_2O adds to its double bond to form malate

4. _____ FAD removes hydrogen to form a double bond

5. _____ formed when the hydroxyl group in malate is oxidizes

6. _____ the compound that is regenerated in the citric acid cycle

7. One turn of the citric acid cycle produces

A. 3 NADH.
B. 3 NADH, 1 $FADH_2$.
C. 3 $FADH_2$, 1 NADH, 1 ATP.
D. 3 NADH, 1 $FADH_2$, 1 ATP.
E. 1 NADH, 1 $FADH_2$, 1 ATP.

8. The citric acid cycle is activated by

A. high ATP levels.
B. NADH.
C. high ADP levels.
D. low ATP levels.
E. succinyl CoA.

9. The end products of the electron transport chain are

A. H_2O + ATP.
B. $CO_2 + H_2O$.
C. $NH_3 + CO_2 + H_2O$.
D. $H_2 + O_2$
E. urea (NH_2CONH_2).

10. How many electron transfers in the electron transport chain provide sufficient energy for ATP synthesis?

A. none
B. 1
C. 2
D. 3
E. 4

11. The electron transport chain

A. produces most of the body's ATP.
B. carries oxygen to the cells.
C. produces $CO_2 + H_2O$.
D. is only involved in the citric acid cycle.
E. operates during fermentation.

In questions 12 through 20, match the components of the electron transport system with the following activities.

A. NAD^+
B. FMN
C. FAD
D. Q
E. cytochromes

12. _____ A mobile carrier that transfers electrons from FMN and $FADH_2$ to cytochromes

13. _____ The coenzyme that accepts hydrogen atoms from NADH

14. _____ Coenzyme derived from niacin

15. _____ The coenzyme used to remove hydrogen atoms from two adjacent carbon atoms to form carbon–carbon double bonds

16. _____ The electron acceptors containing iron

17. _____ A coenzyme derived from quinone

18. _____ Coenzymes that contain flavin

19. _____ The reduced form of this coenzyme generates 3 molecules of ATP.

20. _____ The reduced form of this coenzyme generates 2 molecules of ATP.

Match the components of ATP synthase with each of the following descriptions.

A. F_0
B. F_1
C. (L) site
D. tight (T) site
E. open (O) site

21. _____ consists of the channel for the return of protons to the matrix

22. _____ binds ADP + P_i

23. _____ consists of a center subunit and three subunits that change conformations

24. _____ ADP + $P_i \longrightarrow$ ATP

25. _____ releases ATP from ATP synthase

Indicate the number of ATP's for the equations in questions 21 through 25. Select answers from below.

A. 2 ATP **B.** 3 ATP **C.** 6 ATP **D.** 12 ATP **E.** 24 ATP **F.** 36 ATP

26. one turn of the citric acid cycle (acetyl CoA $\longrightarrow$ $2CO_2$)

27. complete combustion of glucose (glucose + $6O_2 \longrightarrow 6H_2O + 6CO_2$)

28. produced when NADH enters electron transport

29. glycolysis (glucose + $O_2 \longrightarrow$ 2 pyruvate + $2H_2O$)

30. oxidation of 2 pyruvate (2 pyruvate $\longrightarrow$ 2 acetyl CoA + $2CO_2$)

Answers to the Practice Test

1. E	**2.** D	**3.** B	**4.** C	**5.** E
6. E	**7.** D	**8.** D	**9.** B	**10.** D
11. A	**12.** D	**13.** B	**14.** A	**15.** C
16. E	**17.** D	**18.** B, C	**19.** A	**20.** C
21. A	**22.** C	**23.** B	**24.** D	**25.** E
26. D	**27.** F	**28.** B	**29.** C	**30.** C

Answers and Solutions to Selected Text Problems

24.1 Other names for the citric acid cycle are the Krebs cycle and tricarboxlyic acid cycle.

24.3 One turn of the citric acid cycle converts 1 acetyl CoA to $2CO_2$, 3NADH + $3H^+$, $FADH_2$, GTP (ATP), and HS—CoA.

24.5 The reactions in steps 3 and 4 involve oxidative decarboxylation, which reduces the length of the carbon chain by one carbon in each reaction.

24.7 NAD^+ is reduced by the oxidation reactions 3, 4, and 8 of the citric acid cycle.

24.9 In reaction 5, GDP undergoes a direct substrate phosphorylation to yield GTP, which converts ADP to ATP and regenerates GDP for the citric acid cycle.

24.11
a. The six-carbon compounds in the citric acid cycle are citrate and isocitrate.
b. Decarboxylation reactions remove carbon atoms as CO_2, which reduces the number of carbon atoms in a chain (reactions 3 and 4).
c. The one five-carbon compound is α-ketoglutarate.
d. Several reactions are oxidation reactions; isocitrate $\longrightarrow$ α-ketoglutarate; α-ketoglutarate $\longrightarrow$ succinyl CoA; succinate $\longrightarrow$ fumarate; malate $\longrightarrow$ oxaloacetate
e. Secondary alcohols are oxidized in Reactions 3 and 8.

24.13
a. Citrate synthase combines oxaloacetate with acetyl CoA.
b. Succinate dehydrogenase and aconitase convert a carbon–carbon single bond to a double bond.
c. Fumarase adds water to the double bond in fumarate; aconitase adds water to the double bond in iconitate from citrate.

24.15
a. NAD^+ accepts 2H from the oxidative decarboxylation of isocitrate.
b. GDP is phosphorylated in the formation of succinate.
c. FAD accepts 2H when the carbon–carbon single bond in succinate is oxidized to a carbon–carbon double bond in fumarate.

24.17 Isocitrate dehydrogenase and α-ketoglutarate dehydrogenase are allosteric enzymes, which increase or decrease the flow of materials through the citric acid cycle.

24.19 High levels of ADP means there are low levels of ATP. To provide more ATP for the cell, the reaction rate of the citric acid cycle increases.

24.21 The Fe^{3+} is the oxidized form of the iron in cytochrome *c*.

24.23 **a.** The loss of $2H^+$ and $2e^-$ is oxidation. **b.** The gain of $2H^+ + 2e^-$ is reduction.

24.25 NADH and $FADH_2$ produced in glycolysis, oxidation of pyruvate, and the citric acid cycle provide the electrons for electron transport.

24.27 FAD is reduced to $FADH_2$, which provides $2H^+$ and $2e^-$ for coenzyme Q, then cytochrome *b*, and then cytochrome *c*.

24.29 The mobile carrier coenzyme Q (or Q) transfers electrons from complex I to III. It also transfers electrons from complex II to complex III.

24.31 When NADH transfers electrons to FMN in complex I, NAD^+ is produced.

24.33 **a.** $NADH + H^+ + \underline{FMN} \longrightarrow \underline{NAD^+} + FMNH_2$
b. $QH_2 + 2\ Fe^{3+}\ \text{cyt b} \longrightarrow \underline{Q} + \underline{2\ Fe^{2+}\ \text{cyt b}} + 2H^+$

24.35 In oxidative phosphorylation, the energy from the oxidation reactions in the electron transport chain is used to drive ATP synthesis.

24.37 Protons must pass through F_0 channel of ATP synthase to return to the matrix. During the process, energy is released to drive the synthesis of ATP in F_1.

24.39 The oxidation of the reduced coenzymes NADH and $FADH_2$ by the electron transport chain generates energy to drive the synthesis of ATP.

24.41 ATP synthase consists of two protein complexes known as F_0 and F_1.

24.43 The loose (L) site in ATP synthase begins the synthesis of ATP by binding ADP and P_i.

24.45 Glycolysis takes place in the cytoplasm, not in the mitochondria. Because NADH cannot cross the mitochondrial membrane, one ATP is hydrolyzed to transport the electrons from NADH to FAD. The resulting $FADH_2$ produces only 2 ATP for each NADH produced in glycolysis.

24.47 **a.** 3 ATP are produced by the oxidation of NADH in electron transport.
b. 2 ATP are produced in glycolysis when glucose degrades to 2 pyruvate.
c. 6 ATP are produced when 2 pyruvate are oxidized to 2 acetyl CoA and 2 CO_2.
d. 12 ATP are produced in one turn of the citric acid cycle as acetyl CoA is converted to 2 CO_2.

24.49 The oxidation reactions of the citric acid cycle produce a source of reduced coenzymes for the electron transport chain and ATP synthesis.

24.51 The electron transport chain regenerates the oxidized forms of the coenzymes NAD^+ and FAD for use again by the citric acid cycle.

24.53 **a.** Citrate and isocitrate are six-carbon compounds in the citric acid cycle.
b. α-Ketoglutarate is a five-carbon compound.
c. The compounds α-ketoglutarate, succinyl-CoA, and oxaloacetate have keto groups.

24.55 **a.** In Reaction 4, α-ketoglutarate, a five-carbon keto acid, is decarboxylated.
b. In Reactions 1 and 7, double bonds in aconitate and fumarate are hydrated.
c. NAD^+ is reduced in Reactions 3, 4, and 8.
d. In Reactions 3 and 8, a secondary hydroxyl group in isocitrate and malate is oxidized.

24.57 **a.** NAD^+ is the coenzyme for the oxidation of a secondary hydroxyl group in isocitrate to a keto group in α-ketoglutarate.
b. NAD^+ and CoA are needed in the oxidative decarboxylation of α-ketoglutarate to succinyl CoA.

24.59 **a.** High levels of NADH inhibit isocitrate dehydrogenase and α-ketoglutarate dehydrogenase to slow the rate of the citric acid cycle.
b. High levels of ATP inhibit the citric acid cycle.

24.61 **a.** A heme group is found in all the cytochromes (4).
b. FMN (1) contains a ribitol group.

24.63 **a.** CoQ is a mobile carrier.
b. Fe-S clusters are found in complexes I and III, and IV.
c. cyt a_3 is part of complex IV

24.65 **a.** $FADH_2$ is oxidized in complex II: $FADH_2 + Q \longrightarrow FAD + QH_2$
b. Cyt a (Fe^{2+}) is oxidized in complex IV:
Cyt a (Fe^{2+}) + Cyt a_3 (Fe^{3+}) → Cyt a (Fe^{3+}) + Cyt a_3 (Fe^{2+})

24.67 Complete the following by adding the substances that are missing:

a.

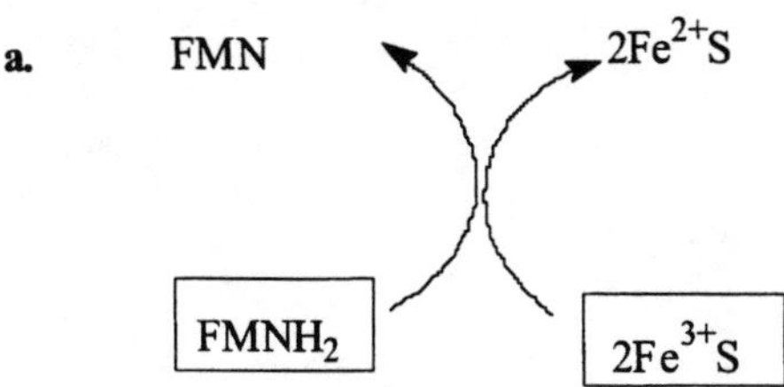

b.

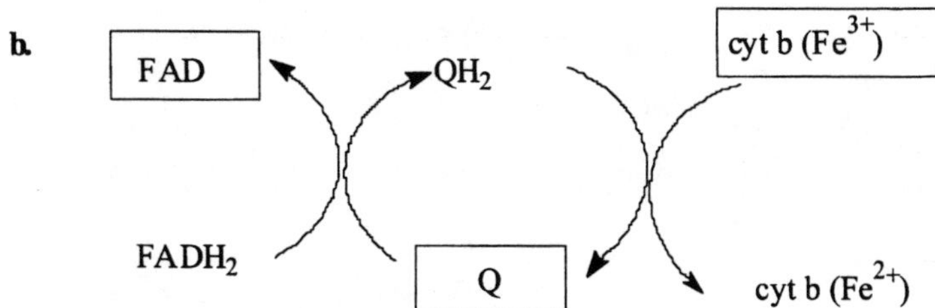

24.69 The transfer of electrons by complexes I, III, and IV generate energy to pump protons out of the matrix into the inner membrane space.

24.71 In the chemiosmotic model, energy released by the flow of protons through the ATP synthase is utilized for the synthesis ATP.

24.73 In the inner membrane space, there is a higher concentration of protons, which reduces the pH and forms an electrochemical gradient. As a result protons flow into the matrix where the proton concentration is lower and the pH is higher.

24.75 Two ATP molecules are produced from the energy generated by the electrons from $FADH_2$ moving through electron transport to oxygen.

24.77 **a.** Amytal and rotenone inhibit the transfer of electrons in NADH dehydrogenase (complex I).
b. Antimycin inhibits electron flow from cyt b to cyt c_1 in complex III.
c. Cyanide and carbon monoxide inhibit the flow of electron through cytochrome c oxidase (complex IV).

24.79 The oxidation of glucose to pyruvate by glycolysis produces 6 ATP. 2 ATP are formed by direct phosphorylation along with 2 NADH. Because the 2 NADH are produced in the cytosol, the electrons are transferred to form 2 $FADH_2$, which produces 4 ATP. The oxidation of glucose to CO_2 and H_2O produces 36 ATP.

24.81 **a.** 6 ATP $\times$ 7.3 kcal/mole = 44 kcal
b. 6 ATP $\times$ 7.3 kcal/mole = 44 kcal (2 pyruvate to 2 acetyl CoA)
c. 24 ATP $\times$ 7.3 kcal/ mole = 175 kcal (2 acetyl CoA citric acid cycle)
d. 36 $\times$ 7.3 kcal/mol = 263 (complete oxidation of glucose to CO_2 and H_2O)

24.83 In a calorimeter, the complete combustion of glucose gives 687 kcal. The efficiency of ATP synthesis is determined by comparing the total kcal in 36 ATP (283 kcal in problem 24.81) to the energy obtained from glucose in a calorimeter. 283/687 X 100 = 38.3% efficient

24.85 The ATP synthase extends through the inner mitochondrial membrane with the F_0 part in contact with the proton gradient in the intermembrane space, while the F_1 complex is in the matrix.

24.87 As protons from the proton gradient move through the ATP synthase to return to the matrix, energy is released and used to drive ATP synthesis at F_1.

24.89 A hibernating bear has stored fat as brown fat, which can be used during the winter for heat rather than ATP energy.

25

Metabolic Pathways for Lipids and Amino Acids

Study Goals

- Describe the sites, enzymes, and products for the digestion of triacylglycerols.
- Describe the oxidation of fatty acids via β-oxidation.
- Calculate the ATP produced by the complete oxidation of a fatty acid.
- Explain ketogenesis and the conditions in the cell that form ketone bodies.
- Describe the biosynthesis of fatty acids from acetyl CoA.
- Describe the sites, enzymes, and products of the digestion of dietary proteins.
- Explain the role of transamination and oxidative deamination in the degradation of amino acids.
- Describe the formation of urea from ammonium ion.
- Explain how carbon atoms from amino acids are prepared to enter the citric acid cycle or other pathways.
- Show how nonessential amino acids are synthesized from substances used in the citric acid cycle and other pathways.

Think About It

1. What are the products from the digestion of triacylalcohols and proteins?

2. What is the purpose of adding the enzyme lactase to milk products?

3. When do you utilize fats and proteins for energy?

Key Terms

Match the key terms with the correct statement shown below:

a. essential amino acid **b.** transamination **c.** lipogenesis
d. ketosis **e.** beta (β)-oxidation

1. _____ A reaction cycle that oxidizes fatty acids by removing acetyl CoA units

2. _____ The synthesis of fatty acids by linking two-carbon acetyl units

3. _____ A condition in which high levels of ketone bodies lower blood pH

4. _____ An amino acid that must be obtained from the diet

5. _____ The transfer of an amino group from an amino acid to an α-keto acid.

Answers **1.** e **2.** c **3.** d **4.** a **5.** b

25.1 Digestion of Triacylglycerols

- Dietary fats begin digestion in the small intestine where they are emulsified by bile salts.
- Pancreatic lipases hydrolyze the triacylglycerols to yield monoacylglycerols and free fatty acids.
- The triacylglycerols reformed in the intestinal lining combine with proteins to form chylomicrons for transport through the lymph system and bloodstream.
- In the cells, triacylglycerols hydrolyze to glycerol and fatty acids, which can be used for energy.

◆ Learning Exercise 25.1

Match each of the following terms with the descriptions that follow.

a. chylomicrons **b.** lipases **c.** fat mobilization
d. monoacylglycerols and fatty acids **e.** emulsification

1. _____ the hydrolysis of triacylglycerols in adipose tissues to produce energy

2. _____ lipoproteins formed when triacylglycerols are coated with proteins

3. _____ the breakup of fat globules in the small intestine by bile salts

4. _____ enzymes released from the pancreas that hydrolyze triacylglycerols

5. _____ the products of lipase hydrolysis of triacylglycerols in the small intestine

Answers **1.** c **2.** a **3.** e **4.** b **5.** d

25.2 Oxidation of Fatty Acids

- When needed for energy, fatty acids link to coenzyme A for transport to the mitochondria where they undergo β-oxidation.
- In β-oxidation, a fatty acyl chain is oxidized to yield a shortened fatty acid, acetyl CoA, and the reduced coenzymes NADH and $FADH_2$.

◆ Learning Exercise 25.2A

Match each of the following terms with the descriptions that follow.

a. activation **b.** fatty acyl carnitine **c.** β-oxidation
d. NAD^+ and FAD **e.** mitochondria

1. _____ coenzymes needed for β-oxidation

2. _____ site in the cell where β-oxidation of fatty acids takes place

3. _____ a fatty acid combines with SH—CoA to form fatty acyl CoA

4. _____ carries the fatty acyl group into the mitochondria matrix

5. _____ the sequential removal of two-carbon sections from a fatty acid

Answers **1.** d **2.** e **3.** a **4.** b **5.** c

◆ Learning Exercise 25.2B

1. Write an equation for the activation of myristic (C_{14}) acid: $CH_3—(CH_2)_{12}—\overset{\overset{\displaystyle O}{\|}}{C}—O^-$

2. Write an equation for the first oxidation of myristyl CoA.

3. Write an equation for the hydration of the double bond.

4. Write the overall equation for the complete oxidation of myristyl CoA.

5. **a.** How many cycles of β-oxidation will be required?
   b. How many acetyl CoA units will be produced?

Answers

1. $CH_3—(CH_2)_{12}—\overset{\overset{\displaystyle O}{\|}}{C}—O^- + HS—CoA + ATP \xrightarrow{\textit{Acyl CoA synthetase}} CH_3—(CH_2)_{12}—\overset{\overset{\displaystyle O}{\|}}{C}—S—CoA + AMP + 2P_i$

2. $CH_3—(CH_2)_{12}—\overset{\overset{\displaystyle O}{\|}}{C}—S—CoA + FAD \xrightarrow{\textit{Acyl CoA dehydrogenase}} CH_3—(CH_2)_{10}—CH{=}CH—\overset{\overset{\displaystyle O}{\|}}{C}—S—CoA + FADH_2$

3. $CH_3—(CH_2)_{10}—CH{=}CH—\overset{\overset{\displaystyle O}{\|}}{C}—S—CoA + H_2O \xrightarrow{\textit{Thiolase}} CH_3—(CH_2)_{10}—\overset{\overset{\displaystyle OH}{|}}{CH}—CH_2—\overset{\overset{\displaystyle O}{\|}}{C}—S—CoA$

4. myristyl (C_{14})—CoA + 6CoA + 6 FAD + 6 NAD^+ + 6 H_2O ⟶
   7Acetyl CoA + 6 $FADH_2$ + 6 NADH + 6H^+

5. **a.** 6 cycles **b.** 7 acetyl CoA units are produced

25.3 ATP and Fatty Acid Oxidation

- The energy obtained from a particular fatty acid depends on the number of carbon atoms.
- Two ATP are required for activation. Then each acetyl CoA produces 12 ATP via citric acid cycle, and the electron transport chain converts each NADH to 3 ATP and each FADH to 2 ATP.

◆ Learning Exercise 25.3

Lauric acid is a 12-carbon fatty acid: $CH_3(CH_2)_{10}COOH$.

1. How much ATP is needed for activation?

2. How many cycles of β-oxidation are required?

3. How many NADH and $FADH_2$ are produced during β-oxidation?

4. How many acetyl CoA units are produced?

5. What is the total ATP produced from the electron transport chain and the citric acid cycle?

Answers **1.** 2 ATP **2.** 5 cycles **3.** 5 cycles produce 5 NADH and 5 $FADH_2$
4. 6 acetyl CoA **5.** 5 NADH × 3 ATP = 15 ATP; 5 $FADH_2$ × 2 ATP = 10 ATP; 6 acetyl CoA × 12 ATP = 72 ATP; Total ATP = 15 ATP + 10 ATP + 72 ATP – 2 ATP (for activation) = 95 ATP

25.4 Ketogenesis and Ketone Bodies

- When the oxidation of large amounts of fatty acids cause high levels of acetyl CoA, the acetyl CoA undergo ketogenesis.
- Two molecules of acetyl CoA form acetoacetyl CoA, which is converted to acetoacetate, β-hydroxybutyrate, and acetone.

◆ Learning Exercise 25.4

Match each of the following terms with the correct description.

a. ketone bodies **b.** ketogenesis **c.** ketosis
d. liver **e.** acidosis

1. _____ high levels of ketone bodies in the blood
2. _____ a metabolic pathway that produces ketone bodies
3. _____ β-hydroxybutyrate, acetoacetate, and acetone
4. _____ the condition whereby ketone bodies lower the blood pH below 7.4
5. _____ site where ketone bodies form

Answers **1.** c **2.** b **3.** a **4.** e **5.** d

25.5 Fatty Acid Synthesis

- When there is an excess of acetyl CoA in the cell, the two-carbon units link together to synthesize fatty acids that are stored in the adipose tissue.
- Two-carbon acetyl CoA units link together to give palmitic (C_{16}) acid and other fatty acids.

◆ Learning Exercise 25.5

Indicate if each of the following are characteristic of lipogenesis (L) or β-oxidation (O).

1. ____ occurs in the matrix of mitochondria

2. ____ occurs in the cytosol of mitochondria

3. ____ activated by insulin

4. ____ activated by glucagon

5. ____ starts with fatty acids

6. ____ starts with acetyl CoA units

7. ____ produces fatty acids

8. ____ produces acetyl CoA units

9. ____ requires NADPH coenzyme

10. ____ requires FAD and NAD^+ coenzymes

11. ____ activated with CoA

12. ____ activated by ACP

Answers **1.** O **2.** L **3.** L **4.** O **5.** O **6.** L
7. L **8.** O **9.** L **10.** O **11.** O **12.** L

25.6 Digestion of Proteins

- Proteins begin digestion in the stomach, where HCl denatures proteins and activates peptidases that hydrolyze peptide bonds.
- In the small intestine trypsin and chymotrypsin complete the hydrolysis of peptides to amino acids.

◆ Learning Exercise 25.6

Match each of the following terms with the correct description.

a. nitrogen-containing compounds **b.** protein turnover **c.** stomach
d. nitrogen balance **e.** small intestine

1. ____ HCl activates enzymes that hydrolyzes peptide bonds in proteins

2. ____ trypsin and chymotrypsin convert peptides to amino acids

3. ____ include amino acids, amino alcohols, protein hormones, and nucleic acids

4. ____ the process of synthesizing protein and breaking them down

5. ____ the amount of protein hydrolyzed is equal to the amount of protein used in the body

Answers **1.** c **2.** e **3.** a **4.** b **5.** d

25.7 Degradation of Amino Acids

- Amino acids are normally used for protein synthesis.
- Amino acids are degraded by transferring an amino group from an amino acid to an α-keto acid to yield a different amino acid and α-keto acid.
- In oxidative deamination, the amino group in glutamate is removed as an ammonium ion, NH_4^+.

◆ Learning Exercise 25.7A

Match each of the following descriptions with transamination (T) or oxidative deamination (D).

1. ____ produces an ammonium ion, NH_4^+
2. ____ transfers an amino group to an α-keto acid
3. ____ usually involves the degradation of glutamate
4. ____ requires NAD^+ or $NADP^+$
5. ____ produces another amino acid and α-keto acid
6. ____ usually produces α-ketoglutarate

Answers

	1. D	2. T	3. D
	4. T	5. T	6. D

◆ Learning Exercise 25.7B

1. Write an equation for the transamination reaction of serine and oxaloacetate.

2. Write an equation for the oxidation deamination of glutamate.

Answers

1. $$HO{-}CH_2{-}\overset{\overset{NH_3^+}{|}}{CH}{-}COO^- + {}^-OOC{-}\overset{\overset{O}{\|}}{C}{-}CH_2{-}COO^- \longrightarrow$$
$$HO{-}CH_2{-}\overset{\overset{O}{\|}}{C}{-}COO^- + {}^-OOC{-}\overset{\overset{NH_3^+}{|}}{CH}{-}CH_2{-}COO^-$$

2. $${}^-OOC{-}\overset{\overset{NH_3^+}{|}}{CH}{-}CH_2{-}CH_2{-}COO^- + NAD^+ \text{ (or } NADP^+\text{)} + H_2O \longrightarrow$$
$${}^-OOC{-}\overset{\overset{O}{\|}}{C}{-}CH_2{-}CH_2{-}COO^- + NH_4^+ + NADH \text{ (or NADPH)} + H^+$$

25.8 Urea Cycle

◆ Learning Exercise 25.8

- The ammonium ion, NH_4^+, from amino acid degradation is toxic if allowed to accumulate.
- The urea cycle converts ammonium ion to urea, which forms urine in the kidneys.

Arrange the following reactions in the order they occur in the urea cycle.

a. argininosuccinate is split to yield arginine and fumarate

b. aspartate condenses with citrulline to yield argininosuccinate

c. arginine is hydrolyzed to yield urea and regenerates ornithine

d. ornithine combines with the carbamoyl group from carbamoyl phosphate

Answers **1.** d **2.** b **3.** a **4.** c

25.9 Fates of the Carbon Atoms from Amino Acids

- α-Keto acids resulting from transamination can be used as intermediates in the citric acid cycle, in the synthesis of lipids or glucose, or oxidized for energy.

◆ Learning Exercise 25.9

Match each of the following terms with the correct description.

a. glucogenic **b.** ketogenic **c.** oxaloacetate
d. acetyl CoA **e.** α-ketoglutarate **f.** pyruvate

1. ____ amino acids that generate pyruvate or oxaloacetate, which can be used to synthesize glucose

2. ____ keto acid obtained from carbon atom of alanine and serine

3. ____ keto acid obtained from carbon atom of glutamine and glutamate

4. ____ keto acid obtained from carbon atoms from aspartate and asparagine

5. ____ amino acids that generate compounds that can produce ketone bodies

6. ____ compound obtained from carbon atoms of leucine and tryptophan

Answers **1.** a **2.** f **3.** e **4.** c **5.** b **6.** d

25.10 Synthesis of Amino Acids

- Humans synthesize only 10 amino acids. The other 10, called essential amino acids, must be obtained from the diet.

◆ Learning Exercise 25.10

Match each of the following terms with the correct description.

a. essential amino acids
b. nonessential amino acids
c. transamination
d. phenylketonurea (PKU)

1. _____ amino acids synthesized in humans

2. _____ a genetic condition when phenylalanine is not converted to tyrosine

3. _____ amino acids that must be supplied by the diet

4. _____ reaction that produces some nonessential amino acids

Answers **1.** b **2.** d **3.** a **4.** c

Check List for Chapter 25

You are ready to take the practice test for Chapter 25. Be sure that you have accomplished the following learning goals for this chapter. If you are not sure, review the section listed at the end of the goal. Then apply your new skills and understanding to the practice test.

After studying Chapter 25 I can successfully:

_____ Describe the sites, enzymes, and products for the digestion of triacylglycerols (25.1).

_____ Describe the oxidation of fatty acids via β-oxidation (25.2).

_____ Calculate the ATP produced by the complete oxidation of a fatty acid (25.3).

_____ Explain ketogenesis and the conditions in the cell that form ketone bodies (25.4).

_____ Describe the biosynthesis of fatty acids from acetyl CoA (25.5).

_____ Describe the sites, enzymes, and products of the digestion of dietary proteins (25.6).

_____ Explain the role of transamination and oxidative deamination in degrading amino acids (25.7).

_____ Describe the formation of urea from ammonium ion (25.8).

_____ Explain how carbon atoms from amino acids are prepared to enter the citric acid cycle or other pathways (25.9).

_____ Show how nonessential amino acids are synthesized from substances used in the citric acid cycle and other pathways (25.10).

Practice Test for Chapter 25

1. The digestion of triacylglycerols takes place in the ____________________ by enzymes called ________________________.

A. small intestine; peptidases
B. stomach; lipases
C. stomach; peptidases
D. small intestine; lipases
E. all of these

2. The products of the digestion of triacylglycerols are

A. fatty acids.
B. monoacylglycerols.
C. glycerol.
D. diacylglycerols.
E. all of these

3. The function of the bile salts in the digestion of fats is

A. emulsification. **B.** hydration. **C.** dehydration.
D. oxidation. **E.** reduction.

4. Chylomicrons formed in the intestinal lining

A. are lipoproteins.
B. are triacylglycerols coated with proteins.
C. transport fats into the lymph system and bloodstream.
D. carry triacylglycerols tp the cells of the heart, muscle, and adipose tissues.
E. all of these

5. Glycerol obtained from hydrolysis enters glycolysis when converted to

A. glucose. **B.** fatty acids.
C. dihydroxyacetone phosphate. **D.** pyruvate. **E.** glycerol-3-phosphate.

6. Fatty acids are prepared for β-oxidation by forming

A. carnitine. **B.** fatty acyl carnitine. **C.** acetyl CoA.
D. fatty acyl CoA. **E.** pyruvate.

7. The reactions in the β-oxidation cycle do *not* involve

A. reduction. **B.** hydration. **C.** dehydrogenation.
D. oxidation. **E.** cleavage of acetyl CoA.

Consider the β-oxidation of palmitic (C_{16}) acid for questions 8 through 11.

8. The number of β-oxidation cycles required for palmitic (C_{16}) acid is

A. 16. **B.** 9. **C.** 8.
D. 7. **E.** 6.

9. The number of acetyl CoA groups produced by the β-oxidation of palmitic (C_{16}) acid is

A. 16. **B.** 9. **C.** 8.
D. 7. **E.** 6.

10. The number of NADH and $FADH_2$ produced by the β-oxidation of palmitic (C_{16}) acid is

A. 16. **B.** 9. **C.** 8.
D. 7. **E.** 6.

11. The total ATP produced by the β-oxidation of palmitic (C_{16}) acid is

A. 96. **B.** 129. **C.** 131.
D. 134. **E.** 136.

12. The oxidation of large amounts of fatty acids can produce

A. ketone bodies. **B.** glucose. **C.** low pH level in the blood.
D. acetone. **E.** pyruvate...

13. The metabolic pathway of lipogenesis requires

A. fatty acids. **B.** acetyl CoA. **C.** FAD and NAD^{+}.
D. glucagon. **E.** ketone bodies.

14. The digestion of proteins takes place in the __________ by enzymes called __________.

A. small intestine; peptidases
B. stomach; lipases
C. stomach; peptidases
D. small intestine; lipases
E. stomach and small intestine; proteases and peptidases

15. The process of transamination

A. is part of the citric acid cycle.
B. converts α-amino acids to β-keto acids.
C. produces new amino acids.
D. is not used in the metabolism of amino acids.
E. is part of the β-oxidation of fats.

16. The oxidative deamination of glutamate produces

A. a new amino acid.
B. a new β-keto acid.
C. ammonia, NH_3.
D. ammonium ion, NH_4^+.
E. urea.

17. The purpose of the urea cycle in the liver is to

A. synthesize urea.
B. convert urea to ammonium ion, NH_4^+.
C. convert ammonium ion NH_4^+ to urea.
D. synthesize new amino acids.
E. is part of the β-oxidation of fats.

18. The urea cycle begins with the conversion of NH_4^+ to

A. asparate.
B. carbamoyl phosphate.
C. citrulline.
D. argininosuccinate.
E. urea.

19. The carbon atoms from a ketogenic amino acid can be used to

A. synthesize ketone bodies.
B. convert urea to ammonium ion, NH_4^+.
C. synthesize fatty acids.
D. produce energy.
E. synthesize proteins.

20. The carbon atoms from various amino acids can be used in several ways such as

A. intermediates of the citric acid cycle
B. formation of pyruvate
C. synthesis of glucose
D. formation of ketone bodies
E. all of these

21. Essential amino acids

A. are not synthesized by humans
B. are required in the diet
C. are excreted if in excess
D. include leucine, lysine, and valine
E. all of these

22. When the quantity of amino acids in the diet exceeds the needs of the cells, the excess amino acids

A. are stored for use at a later time
B. are used to synthesize glycogen
C. excreted
D. are converted to fat
E. are used to make more protein

23. Phenylketonuria is a condition

A. abbreviated as PKU
B. where a person does not synthesize tyrosine
C. that can be detected at birth
D. causes severe mental retardation
E. all of these

Answers to the Practice Test

1. D	2. E	3. A	4. E	5. C
6. D	7. A	8. D	9. C	10. D
11. B	12. A	13. A	14. E	15. C
16. D	17. C	18. B	19. A	20. E
21. E	22. C	23. E		

Answers and Solutions to Selected Text Problems

25.1 The bile salts emulsify fat to give small fat globules for lipase hydrolysis.

25.3 Fats are mobilized when blood glucose and glycogen stores are depleted.

25.5 Glycerol is converted to glycerol-3-phosphate and then to dihydroxyacetone phosphate, which is an intermediate of glycolysis.

25.7 Fatty acids are activated in the cytosol of the mitochondria.

25.9 The coenzymes FAD and NAD^+ are required for β-oxidation.

25.11 The designation β carbon is based on the common names of carboxylic acids whereby the alpha β carbon and the β carbons are adjacent to the carboxyl group.

a. $CH_3—CH_2—CH_2—CH_2—CH_2—\overset{\beta}{C}H_2—CH_2—\overset{O}{\overset{\|}{C}}—S—CoA$

b. $CH_3—(CH_2)_{14}—\overset{\beta}{C}H_2—CH_2—\overset{O}{\overset{\|}{C}}—S—CoA$

c. $CH_3—CH_2—CH{=}CH—CH_2—CH_2—CH_2—\overset{\beta}{C}H_2—CH_2—\overset{O}{\overset{\|}{C}}—S—CoA$

25.13 **a., b.** $CH_3—(CH_2)_6—\underset{\beta}{CH_2}—\underset{\alpha}{CH_2}—\overset{O}{\overset{\|}{C}}—S—CoA$

c. $CH_3—(CH_2)_8—\overset{O}{\overset{\|}{C}}—S—CoA + NAD^+ + FAD + H_2O + SH—CoA \longrightarrow$

$CH_3—(CH_2)_6—\overset{O}{\overset{\|}{C}}—S—CoA + CH_2—\overset{O}{\overset{\|}{C}}—S—CoA + NADH + H^+ + FADH_2$

d. $CH_3—(CH_2)_8—COOH + 4CoA + 4FAD + 4NAD^+ + 4H_2O \longrightarrow$
$5Acetyl\ CoA + 4FADH_2 + 4NADH + 4H^+$

25.15 The hydrolysis of ATP to AMP hydrolyzes ATP to ADP, and ADP to AMP, which provides the same amount of energy as the hydrolysis of 2 ATP to 2 ADP.

25.17 **a.** The β-oxidation of a chain of 10 carbon atoms produces 5 acetyl CoA units.
b. A C_{10} fatty acid will go through 4 β-oxidation cycles.
c. 60 ATP from 5 acetyl CoA (citric acid cycle) + 12 ATP from 4 NADH + 8 ATP from 4 $FADH_2$ –2 ATP (activation) = 80 –2 = 78 ATP

25.19 Ketogenesis is the synthesis of ketone bodies from excess acetyl CoA from fatty acid oxidation, which occurs when glucose is not available for energy. This occurs in starvation, fasting, and diabetes.

25.21 Acetoacetate undergoes reduction using NADH + H^+ to yield β-hydroxybutyrate.

25.23 High levels of ketone bodies lead to ketosis, a condition characterized by acidosis (a drop in blood pH values), and characterized by excessive urination and strong thirst.

25.25 Fatty acid synthesis takes place in the cytosol of cells in liver and adipose tissue.

25.27 Fatty acid synthesis starts when acetyl CoA, HCO_3^-, and ATP produce malonyl CoA.

25.29
a. (3) malonyl CoA transacylase converts malonyl CoA to malonyl ACP.
b. (1) acetyl CoA carboxylase combines acetyl CoA with bicarbonate to yield malonyl CoA.
c. (2) acetyl CoA transacylase converts acetyl CoA to acetyl ACP.

25.31 Capric acid is a fatty acid with 10 carbon atoms, $C_{10}H_{20}O_2$. How are each of the following involved in the synthesis of a molecule of capric acid?

a. A C_{10} fatty acid requires the formation of 4 malonyl ACP, which uses 4 HCO_3^-.
b. 4 ATP are required to produce 4 malonyl CoA.
c. 5 acetyl CoA are needed to make 1 acetyl ACP and 4 malonyl ACP.
d. A C_{10} fatty acid requires 4 malonyl ACP and 1 acetyl ACP.
e. A C_{10} fatty acid chain requires 4 cycles with 2 NADPH/cycle or a total of 8 NADPH.
f. The four cycles remove a total of 4 CO_2.

25.33 The digestion of proteins begins in the stomach and is completed in the small intestine.

25.35 Nitrogen-containing compounds in the cells include hormones, heme, purines and pyrimidines for nucleotides, proteins, nonessential amino acids, amino alcohols, and neurotransmitters.

25.37 The reactants are an amino acid and an α-keto acid, and the products are a new amino acid and a new α-keto acid.

25.39 In transamination, an amino group replaces a keto group in the corresponding α-keto acid.

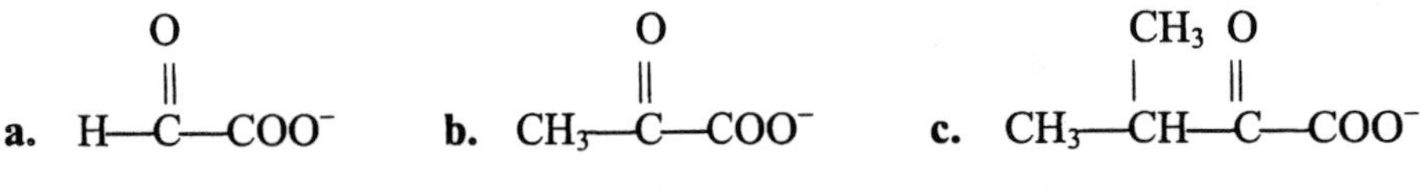

a. $H{-}\overset{\overset{\large O}{\|}}{C}{-}COO^-$ **b.** $CH_3{-}\overset{\overset{\large O}{\|}}{C}{-}COO^-$ **c.** $CH_3{-}\overset{\overset{\large CH_3}{|}}{CH}{-}\overset{\overset{\large O}{\|}}{C}{-}COO^-$

25.41 In an oxidative deamination, the amino group in an amino acid such as glutamate is removed as an ammonium ion. The reaction requires NAD^+ or $NADP^+$.

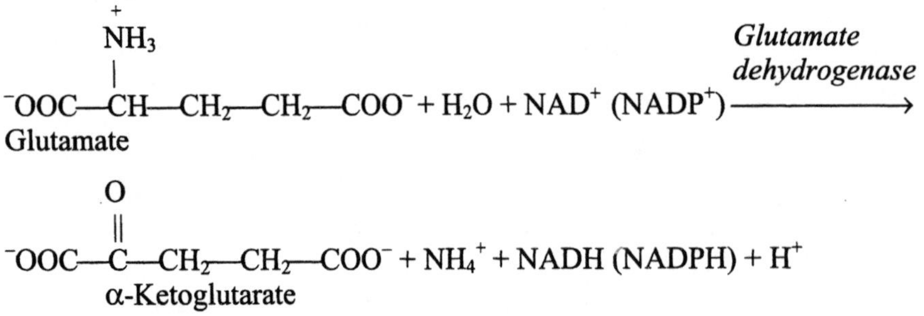

$$^-OOC{-}\overset{\overset{\large \overset{+}{N}H_3}{|}}{CH}{-}CH_2{-}CH_2{-}COO^- + H_2O + NAD^+\ (NADP^+) \xrightarrow{\textit{Glutamate dehydrogenase}}$$
Glutamate

$$^-OOC{-}\overset{\overset{\large O}{\|}}{C}{-}CH_2{-}CH_2{-}COO^- + NH_4^+ + NADH\ (NADPH) + H^+$$
α-Ketoglutarate

25.43 NH_4^+ is toxic if allowed to accumulate in the liver.

25.45 $H_2N—\overset{\overset{\displaystyle O}{\|}}{C}—NH_2$

25.47 The carbon atom in urea is obtained from the CO_2 produced by the citric acid cycle.

25.49 Glucogenic amino acids can be used to produce intermediates for glucogenesis, which is glucose synthesis.

25.51 **a.** The three-carbon atom structure of alanine is converted to pyruvate.
b. The four-carbon structure of aspartate is converted to fumarate or oxaloacetate.
c. Valine is converted to succinyl CoA.
d. The five-carbon structure from glutamine can be converted to α-ketoglutarate.

25.53 Humans can synthesize only nonessential amino acids.

25.55 Glutamine synthetase catalyzes the addition of an amino group to glutamate using energy from the hydrolysis of ATP.

25.57 Phenylketonurnia

25.59 Triacylglycerols are hydrolyzed to monoacylglycerols and fatty acids in the small intestine, which are reformed into triacylglycerols in the intestinal lining for transport as lipoproteins to the tissues.

25.61 Fats can be stored in unlimited amounts in adipose tissue compared to the limited storage of carbohydrates as glycogen.

25.63 The fatty acids cannot diffuse across the blood–brain barrier.

25.65 **a.** Glycerol is converted to glycerol-3-phosphate and to dihydroxyacetone phosphate, which can enter glycolysis or gluconeogenesis.
b. Activation of fatty acids occurs on the outer mitochondrial membrane.
c. The energy cost is equal to 2 ATP.
d. Only fatty acyl CoA can move into the intermembrane space for transport by carnitine into the matrix.

25.67 Lauric acid, $CH_3—(CH_2)_{10}—COOH$, is a C_{12} fatty acid. ($C_{12}H_{24}O_2$)

a. and **b.** $CH_3—(CH_2)_8—\underset{\beta}{CH_2}—\underset{\alpha}{CH_2}—\overset{\overset{\displaystyle O}{\|}}{C}—CoA$

c. Lauryl-CoA + 5 CoA + 5 FAD + 5 NAD^+ + 5 H_2O $\longrightarrow$
6 Acetyl CoA + 5 $FADH_2$ + 5 NADH + 5H^+

d. Six acetyl CoA units are produced.

e. Five cycles of β oxidation are needed.

f.

	activation	$\longrightarrow$	−2 ATP
6	acetyl CoA × 12	$\longrightarrow$	72 ATP
5	$FADH_2$ × 2	$\longrightarrow$	10 ATP
5	NADH × 3	$\longrightarrow$	15 ATP
		Total	95 ATP

25.69 **a.** β oxidation **b.** β oxidation
c. Fatty acid synthesis **d.** β oxidation
e. Fatty acid synthesis **f.** Fatty acid synthesis

25.71 **a.** (1) fatty acid oxidation **b.** (2) the synthesis of fatty acids

25.73 Ammonium ion is toxic if allowed to accumulate in the liver.

25.75 **a.** Citrulline **b.** Carbamoyl phosphate

25.77 **a.** The carbon atom structure of valine is converted to pyruvate.
b. Isoleucine, a ketogenic amino acid, is converted to succinyl CoA or acetyl CoA.
c. The degradation of methionine produces succinyl-CoA.
d. Glutamate can be converted to five-carbon α-ketoglutarate.

25.79 Serine is degraded to pyruvate, which is oxidized to acetyl CoA. The oxidation produces NADH + H^+, which provides 3 ATP. In one turn of the citric acid cycle, the acetyl CoA provides 12 ATP. Thus, serine can provide a total of 15 ATP.